Topics in Context

Science, Technology and the Environment

Cornelsen

Topics in Context   Science, Technology and the Environment

Im Auftrag des Verlages herausgegeben von
Prof. Hellmut Schwarz, Mannheim

Erarbeitet von
Dr. Paul Maloney, Hildesheim
Oliver Meyer, Buxheim
sowie
Sabine Otto, Halle; Angela Ringel-Eichinger, Bietigheim-Bissingen;
Geoff Sammon, Bonn

In Zusammenarbeit mit der Englischredaktion
Hartmut Tschepe (verantwortlicher Redakteur);
Dr. Ilka Soennecken (Bildredaktion); Cornelia Hansch (Redaktionsleitung)

Design und technische Umsetzung
Petra Eberhard Grafik Design, Berlin

Gestaltung des Lead-ins
Anette Schamuhn, Berlin

Umschlaggestaltung
Klein & Halm Grafikdesign, Berlin

Dieses Themenheft wurde auf der Grundlage des Kapitels *Science, Technology and the Environment* des Oberstufenlehrwerks *Context 21* entwickelt. Das *Teacher's Manual* (ISBN 978-3-06-032907-6) enthält eine Audio-CD und eine DVD-ROM/Video (mit den Videos zum Schülerheft, interaktiven Tafelbildern für Whiteboard und Beamer, Vorschlägen zur Leistungsmessung und weiteren Materialien).

www.cornelsen.de

Die Webseiten Dritter, deren Internetadressen in diesem Lehrwerk angegeben sind, wurden vor Drucklegung sorgfältig geprüft. Der Verlag übernimmt keine Gewähr für die Aktualität und den Inhalt dieser Seiten oder solcher, die mit ihnen verlinkt sind.

1. Auflage, 5. Druck 2024

Alle Drucke dieser Auflage sind inhaltlich unverändert und können im Unterricht nebeneinander verwendet werden.

© 2011 Cornelsen Verlag, Berlin
© 2018 Cornelsen Verlag GmbH, Mecklenburgische Str. 53, 14197 Berlin

Druck: H. Heenemann, Berlin

ISBN 978-3-06-032887-1

PEFC zertifiziert
Dieses Produkt stammt aus nachhaltig bewirtschafteten Wäldern und kontrollierten Quellen.
www.pefc.de

PEFC
PEFC/04-31-1156

Contents

abbr	abbreviation		
adj	adjective		
adv	adverb		
AE	American English		
BE	British English		
cf.	confer, see		
e.g.	(Latin) exempli gratia = for example		
esp.	especially		
et al.	(Latin) et alii/alia = and other people or things		
etc.	(Latin) et cetera = and so on		
ff.	and the following lines/pages		
fig	figurative(ly)		
fml	formal		
i.e.	(Latin) id est = that is, in other words		
infml	informal		
jdm./ jdn.	jemandem/ jemanden		
l./ll.	line/lines		
n	noun		
p./pp.	page/pages		
pl	plural		
sb.	somebody		
sing	singular		
sl	slang		
sth.	something		
usu.	usually		
v	verb		

CD 02 indicates that the listening text(s) can be found on the audio-CD in the Teacher's Manual (Track 2).

DVD indicates that the video(s) can be found on the DVD-ROM/video in the Teacher's Manual.

IWB indicates that interactive material (for use with an interactive whiteboard or a projector) can be found on the DVD-ROM/video in the Teacher's Manual.

EXTRA indicates additional (optional) materials and tasks.

Webcode: TOP328871-11

is a code that can be entered at www.cornelsen.de/webcode. This connects you directly to a specific website related to a section of this book.

> **Webcode:**
> TOP328871-11

OECD* indicates that the word, expression or name (here: *OECD*) is explained in the Glossary on pp. 52–55.

☆ indicates that the American English pronunciation follows.

Science, Technology and the Environment

'Nuclear waste problem? I don't know about you, but I don't want my kids growing up in a world where there aren't any problems left to solve!'

Language help 1

- the consequences of ...
- meet the challenge of ...
- completely irresponsible
- the disposal of radioactive waste
- leave it to the next generation
- interested in short-term profits
- show no regard for ...

Language help 2

- ... was invented by ...
- ... originally developed for ...
- ... made it possible to ...
- ... led to a radical change in ...
- ... greatly enhanced our ability to ...
- ... have negative consequences for the environment

1 Talking about the cartoon

a Examine the cartoon above closely. In a few sentences, describe the situation: What can you see in the picture? What do you think has just happened?

b Working together with a partner, write the dialogue between the two men in which the sentence in the cartoon occurs. Try and bring the dialogue to some kind of conclusion (e.g. complete agreement or disagreement).
- **Language help 1**

c Read your dialogues aloud and compare your ideas about the conflict expressed in the cartoon.

d Analyse the cartoon and its message.

2 **IWB** **Talking about the photos**

Choose one of the six photos on the right-hand page. Find three or four other students in your class who have chosen the same photo. Together, answer the following questions:
- What does the picture show? In what field(s) is the object used?
- What do you know about its history and/or development?
- How has it changed our lives for the better or for the worse?
- **Language help 2**

3 **IWB** Speculating about the future

a What do you think the world will be like in 30 years' time?
Decide which of the following statements you agree with:

1 Most people on our planet will live in peace and prosperity.
2 Hard or unpleasant labour will be done by robots.
3 People will live longer and have healthier lives.
4 Travel to outer space will be common.
5 Abundant energy will be available from renewable* sources.
6 Climate change will no longer be a problem.

b Compare your answers with a partner, and discuss the points on which you disagree. ▪ **Language help**
c Together with your partner, make two more statements about life in the future.
d Swap statements with another pair. Decide whether or not you agree. Then discuss your reactions to all four statements in a group of four.

Language help

- What makes you think that ...?
- Why are you so sure that ...?
- I don't agree with you on that point, because ...
- Do you really think it's likely that ...?
- I'm quite convinced that the problem of ... can be solved.
- If progress continues at the same rate, ...

Living in Wonderland: Modern Technology

A1 Getting Connected Thomas L. Friedman[*]

- Before you read the following text, make a list of all the different ways in which you communicate with other people. Rank them according to importance. Discuss why some forms of communication are more important to you than others.

I arrived at Paris Charles de Gaulle Airport the other night and was met by a driver sent by a French friend. The driver was carrying a sign with my name on it, but as I approached him I noticed that he was talking to himself, very animatedly. As I got closer, I realized he had one of those Bluetooth wireless phones clipped to his ear and was deep in conversation. I pointed at myself as the person he was supposed to meet. He nodded and went on talking to whomever was on the other end of his phone. 5 ... 10

When my luggage arrived, I grabbed it off the belt; he pointed toward the exit and I followed, as he kept talking on his phone. When we got into the car, I said, 'Do you know my hotel?' He said, 'No.' I showed him the address, and he went back to talking on the phone. 15

After the car started to roll, I saw he had a movie playing on the screen in the dashboard – on the flat panel that usually displays the G.P.S. road map. I noticed this because between his talking on the phone and the movie, I could barely concentrate. I, alas, was in the back seat trying to finish a column on my laptop. When I wrote all that I could, I got out my iPod and listened to a Stevie Nicks album, while he went on talking, driving and watching the movie. 20

After I arrived at my hotel, I reflected on our trip: The driver and I had been together for an hour, and between the two of us we had been doing six different things. He was driving, talking on his phone and watching a video. I was riding, working on my laptop and listening to my iPod. 25

There was only one thing we never did: Talk to each other. [...]

I relate all this because it illustrates something I've been feeling more and more lately – that technology is dividing us as much as uniting us. Yes, technology can make the far feel near. But it can also make the near feel very far. For all I know, my driver was talking to his parents in Africa. How wonderful! But that meant the two of us wouldn't talk at all. And we were sitting two feet from each other. 30

When I shared this story with Linda Stone, the technologist who once labeled the disease of the Internet age 'continuous partial attention' – two people doing six things, devoting only partial attention to each one – she remarked: 'We're so accessible, we're inaccessible. We can't find the off switch on our devices or on ourselves. ... We want to wear an iPod as much to listen to our own playlists as to block out the rest of the world and protect ourselves from all that noise. We are everywhere – except where we actually are physically.' 35

From: 'The Taxi Driver', *The New York Times*, 1 November 2006

4 **approach sb./sth.** come near to sb./sth.
5 **animated** full of interest and energy
17 **dashboard** Armaturenbrett
panel ['pænl] screen
18 **barely** hardly
19 **alas** *(old use)* expression of regret
column Kolumne (Zeitungsartikel)
22 **reflect on sth.** think about sth.
27 **relate sth.** tell sth. (e.g. a story)
33 **continuous** happening without interruption
34 **devote** widmen

1 Comprehension

a *Partner A:* You are a police inspector who has mistaken Friedman for a criminal with a similar name. Ask Thomas Friedman what he did and what he observed from the time he left the plane in Paris to his arrival at his hotel. Be exact – you want to know all the details.

Partner B: You are Thomas Friedman. A police inspector is asking you questions about your taxi ride after landing in Paris. Answer as precisely as you can. After every three questions, swap roles.

b List the devices mentioned in the text and say what they are used for. Which of them help communication in some form, which hinder communication?

c Divide the text into sections. Summarize the contents of each section in one sentence, using a phrase from the box on the right.

> quote an opinion ·
> relate an incident or story ·
> comment on a trend or observation

2 Looking at the language

In the first part of his text, Friedman often describes actions that accompany some other action:

'as I approached him I noticed that *he was talking to himself*' (ll. 4–5)
'He nodded and *went on talking*' (l. 9)

Which structures does the author use here? Find further examples of each in the text. Explain what effect they produce.

3 Stylistic devices*

The author uses several common stylistic devices in his text: paradox, antithesis, juxtaposition.

Examine these four quotations from the text; express each of them in your own words. Decide which device is used and what effect it produces in its context.

▶ Skill 2: Paraphrasing (p. 31)

1 Yes, technology can make the far feel near. But it can also make the near feel very far. (ll. 28–29)
2 But that meant the two of us wouldn't talk at all. And we were sitting two feet from each other. (ll. 30–31)
3 We're so accessible, we're inaccessible. (ll. 34–35)
4 We are everywhere – except where we actually are physically. (ll. 37–38)

4 Beyond the text

a Describe situations you have experienced that illustrate 'continuous partial attention' (l. 33). What role did technology play?

b Discuss Linda Stone's statement that continuous partial attention is 'the disease of the Internet age' (ll. 32–33). Do you agree?

c Comment on Friedman's remark that 'technology is dividing us as much as uniting us' (l. 28), referring also to the photograph on the right.

▶ Further Practice 1–2 (p. 26)

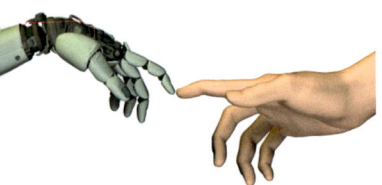

A2 In Our Image: The Age of Robotics

- Describe robots you have seen in science-fiction films. What did they look like? How did they behave? The picture on the left should remind you of a famous painting. What did the artist want to suggest about the relationship between man and machine?

▶ Word Help (p. 49)

Isaac Asimov

i) CD 02 The Three Laws of Robotics Isaac Asimov[*]

Isaac Asimov (1920–1992) was one of America's most successful authors of science-fiction and science books in the 20th century. His novels and short stories depicting the interaction of robots and human beings (e.g. I, Robot, 1950) had a tremendous impact on the genre. Listen to the interview about Asimov's ideas on robots and robotics.

1 While listening

Some of these statements about the interview are incorrect. Write down the numbers of the incorrect sentences while you listen.

1 The word 'robotics' was invented by Asimov in 1942.
2 Asimov read about the *Three Laws of Robotics* in a short story.
3 Asimov is considered the discoverer of the positron.
4 Asimov created the idea of a humanoid machine made of metal.
5 Asimov developed the idea of building safeguards into robot design.
6 Asimov thinks that science-fiction films are far ahead of printed science fiction.

2 After listening

a Correct the incorrect statements in **1**.
b Write down the *Three Laws of Robotics* from memory. Then compare your version with your partner's. Discuss where you might be wrong.

3 Talking about the author's ideas

How would an 'Asimov robot' behave in each of the following situations? Use modal auxiliaries to explain why the robot reacts as it does.

1 A robot sees a child in a burning house.
2 It is instructed to go into a building and detonate a bomb.
3 It sees a man being attacked by another man.

Cynthia Breazeal

ii) Designing Robots for People Cynthia Breazeal[*]

In the following excerpt, MIT engineer Cynthia Breazeal explains how robots should be designed for interaction with human beings.

The next big thing is how you would interact with people. [...] And if you're talking about robots in human society, you're not talking about specialists, you're talking about the average layperson – Grandma, children, and people who know nothing about robotics, now interacting with these kinds of machines. That presented a whole set of challenges. I realized that the human environment is a profoundly social environment, and these robots are going to have to be able to do things not just independently of people but work with people, communicate with

[6] **profound** very great; felt or experienced very strongly

5

people, really be an integrated part of people's lives. Suddenly the emotional intelligence was very, very important because people are going to try to interact with
10 these robots not as tools but as other animate life-like things. If you wanted to make the most natural interface possible, people are already pretty much experts at emotional interaction. The idea was to try to design robots that supported what we were already really good at rather than forcing people to learn a bizarre interface or a bizarre way of communicating to the robot.

15 There's an entirely different range of applications these kinds of robots could serve. So before it was like sweeping minefields and going to explore the ocean, now it's about doing things with people, helping people. In Japan one of the biggest applications or motivations for wanting to develop these emotionally interactive, intelligent robots is that they're concerned about their growing elderly population
20 and the fact that soon they're not going to have enough young people to really tend to all the elder people. So they see robots as a positive technology to help the elderly live independently longer.

The elderly are often reticent about picking up a new technology, so it can't be something too confusing or esoteric. It probably has to be something that they see
25 as genuinely helpful, but in the big picture people should actually really enjoy having these robots around as well. In many ways I think about a blind person's relationship with a seeing eye dog. The seeing eye dog performs a very critical function for that person, a very pragmatic, useful function. But on the other hand, people adore having their dog! So my vision was to use this social form of interaction
30 to really address the needs of a person on a holistic level, not just about helping them with their cognitive and physical abilities, but also appreciate that people are social and emotional creatures and they have pleasure in interacting with things in this way.

From: 'Robot Pals, A Conversation with Cynthia Breazeal', website of PBS, 1 March 2005

10 animate ['ænɪmət] living, having life
11 interface Schnittstelle
15 range Spektrum, Bandbreite
application the practical use of sth., esp. a theory or discovery
23 reticent ['retɪsənt] shy, reserved
29 adore sb./sth. love sb./sth. very much
30 holistic considering a whole thing or being to be more than a collection of parts
31 cognitive ['kɒgnətɪv] connected with mental processes of understanding
appreciate [ə'pri:ʃieɪt] understand that sth. is true, recognize sth.

1 Summarizing the text
a Write a heading for each of the three paragraphs of the text.
b Make notes for each of the paragraphs of the text.
c Summarize Breazeal's ideas on robot design, explaining how she arrived at them.

▶ Skill 11: Writing a summary (p. 38)

2 Language work
a The words in the box on the right are all in the text. Choose the word that means:

1 someone who isn't an expert or professional
2 reasons for doing something
3 strange, unfamiliar
4 truly, in fact
5 practical, down-to-earth

▶ Skill 2: Paraphrasing (p. 31)

motivation · genuinely · layperson · pragmatic · bizarre

b Use each of the words in a sentence of your own.

3 Beyond the text
Working in a group of three, develop a further application in which robots could help solve human problems. What skills would the robot need? What obstacles would have to be overcome? Draw a picture or a diagram of your robot, label it, and present your ideas to your class.

 **Webcode:** TOP328871–11

▶ Further Practice 3 (p. 26)

The Scientific Revolution

The last century saw science and technology emerge as one of the driving forces of modern civilization. The experimental method, based on hypothesis, close observation, and the collection of empirical data, has led to a veritable revolution in our way of life. While pure science engages in basic research, pushing back the frontiers of human knowledge, the applied sciences and engineering concentrate on implementing new discoveries. These have resulted in major advances in communication, transportation and many other fields. $\qquad$ 5

Breakthroughs in medicine and physics, for example, have made it possible for people to live longer and healthier lives than ever before, but also to destroy all life on the planet at the push of a button. Yet few people doubt the necessity of continued $\quad$ 10 progress, and the use of state-of-the-art technology is considered essential for success in the global economy.

In recent years, the life sciences, especially genetics and biotechnology, have come under fire for ethical reasons. Critics of genetically modified (GM) food, for example, often call for a moratorium on the deployment of GM food as long as the long-term $\quad$ 15 effects have not been sufficiently investigated. Introducing artificially created plants into the environment could have unforeseen consequences, they warn.

The controversy surrounding stem cell research and cloning* is based on more general moral objections: it is felt to be wrong for humans to tamper with nature and to destroy human embryos as part of the process. $\qquad$ 20

One of the major challenges facing us in the 21st century is striking a balance between economy and ecology. As global warming, the gradual depletion of natural resources, and the loss of biodiversity threaten the existence of life on our planet, environmentalists are calling for a radical change in our attitudes. To quantify the environmental impact of human activities, they have invented the term 'carbon $\quad$ 25 footprint': the larger the carbon footprint, the more a certain activity contributes to the greenhouse effect. Sustainability has become the keyword of the new century: the reduction of greenhouse gases, the implementation of renewable* energy resources (e.g. solar and wind energy), and conservation of our natural resources have come to be regarded as essential for the survival of the human species. It $\quad$ 30 remains to be seen whether science, having profoundly changed life on this planet, will successfully rise to the challenge of dealing with the consequences of our technical civilization.

1 Word formation and word families

Find the missing word; use a dictionary if you need help.

▶ Skill 3: Using a dictionary (p.32)

1	verb:	*modify*	noun:	? ? ? ? ? ? ? ? ? ?
2	noun:	*deployment*	verb:	? ? ? ? ? ? ? ? ? ?
3	noun:	*controversy*	adjective:	? ? ? ? ? ? ? ? ? ?
4	noun:	*objection*	verb:	? ? ? ? ? ? ? ? ? ?
5	noun:	*depletion*	verb:	? ? ? ? ? ? ? ? ? ?
6	noun:	*sustainability*	adjective:	? ? ? ? ? ? ? ? ? ?
7	adjective:	*renewable*	noun:	? ? ? ? ? ? ? ? ? ?
8	noun:	*conservation*	verb:	? ? ? ? ? ? ? ? ? ?

2 Connotations

Group the words in the box on the right into two lists, according to whether they have a positive or a negative connotation. Discuss the words you are not sure about. The way they are used in the text should also give you a clue to the connotation.

progress · unforeseen · tamper · challenge · depletion · threaten

3 EXTRA Activate your vocabulary

Translate the following sentences. Pay special attention to the underlined expressions that are highlighted in the text.

▶ Skill 5: Translating (p.33)

1 Im 20. Jahrhundert hat die Grundlagenforschung zu mehreren Durchbrüchen in der Naturwissenschaft geführt.
2 In den USA und Europa sind vor allem große Biotechnologie-Konzerne ins Kreuzfeuer der Kritik geraten.
3 Vor dem Einsatz genetisch veränderter Pflanzen sind Untersuchungen der Langzeitwirkungen erforderlich.
4 Mit dem Ausdruck „carbon footprint" quantifiziert man die Auswirkung unseres Handelns auf die Umwelt.

4 Words in connection

a Complete the expressions with the missing preposition (or other linking word). Consult the text if necessary.

1 the study was based ... data collected in the 1990s
2 scientists who engage ... basic research
3 a moratorium ... further testing
4 the controversy ... GM food
5 environmentalists are calling ... a new approach
6 we must all rise ... the challenge of preventing environmental disaster

b Match the words in the two lists to form collocations like those in the text. Then use each of the phrases in a sentence of your own.

strike	a balance
driving	resources
push back	attitude
a change in	force
natural	the frontier

▶ Further Practice 4–6 (p.27)

Cracking the Code: Genetics

B1 How Designer Children Will Work

- First, match the terms on the left with the definitions on the right.

1	gene	a	part of a living cell that contains genes
2	genome	b	organic chemical in which genetic information is encoded
3	chromosome	c	technique for making genetically identical copies of an organism
4	uterus		
5	in vitro	d	done outside the body, i.e. in a laboratory
6	cloning*	e	the smallest unit of hereditary information
7	DNA	f	the complete set of genes belonging to an organism
8	nucleus	g	cell that is capable of developing into one of several types
9	genetic engineering*	h	organ in a woman's body in which an embryo develops
		i	thread-like structure that carries genes
10	stem cell	j	technique of artificially manipulating the genetic structure of an organism

- **hereditary** [hə'redɪtri] given to a child by its parents before it is born

▶ Skill 2: Paraphrasing (p. 31)

The diagram below explains how a procedure known as embryo screening* or PGD (pre-implantation genetic diagnosis*) works.

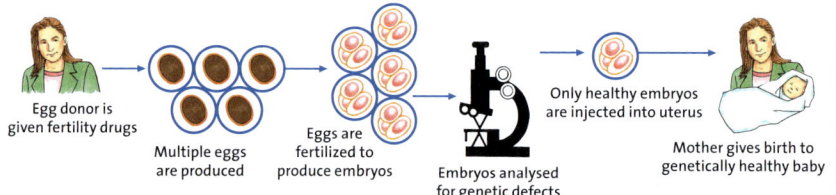

Egg donor is given fertility drugs — Multiple eggs are produced — Eggs are fertilized to produce embryos — Embryos analysed for genetic defects — Only healthy embryos are injected into uterus — Mother gives birth to genetically healthy baby

Language help

- after removal from …
- through artificial insemination in vitro
- under the microscope
- embryos free of genetic defects
- the selected embryo

▶ Word Help (pp. 49–50)

1 Analysing a diagram

a Use the diagram to explain PGD to a partner. Discuss any points that are unclear to you. ● **Language help**

b Write a short article explaining how PGD can be used to prevent children being born with specific genetic defects.

2 **DVD** Collecting information from a film

a Now watch the video about a couple who decided to use embryo screening before the birth of their child. One half of the class takes notes on the benefits connected with this procedure, the other half concentrates on the problems.

b Compare your notes with someone from your group. Add any information that is missing in your notes.

c Now find someone who dealt with the other aspect. Compare your information by discussing each of the following points:

- why the Kingsburys were worried about their future children's health
- how great the risk of dying of cancer is for persons with the same gene mutation as Chad Kingsbury
- how the Kingsburys' family and friends reacted to their decision
- who paid for the necessary treatment
- how the Kingsburys feel about their decision today, after Chloe's birth.

B2 **Born for a Purpose** Jodi Picoult*

- List all the different reasons why people might want to have children.

In Jodi Picoult's novel My Sister's Keeper *(2004), the 13-year-old Anna, whose older sister Kate suffers from leukaemia, thinks about this question.*

Whesn I was little, the great mystery to me wasn't *how* babies were made, but why. The mechanics I understood – my older brother Jesse had filled me in – although at the time I was sure he'd heard half of it wrong. Other kids my age were busy
5 looking up the words penis and vagina in the classroom dictionary when the teacher had her back turned, but I paid attention to different details. Like why some mothers only had one child, while other families seemed to multiply before your eyes. Or how the new girl in school, Sedona, told anyone who'd listen that she was named for
10 the place where her parents were vacationing when they made her. (*'Good thing they weren't staying in Jersey City,'* my father used to say).

Now that I am thirteen, these distinctions are only more complicated: the eighth-grader who dropped out of school because she *got into trouble*; a neighbour who *got herself pregnant* in the hopes it would keep her husband from filing for
15 divorce. I'm telling you, if aliens landed on Earth today and took a good hard look at why babies get born, they'd conclude that most people have children by accident, or because they drink too much on a certain night, or because birth control isn't one hundred percent, or for a thousand other reasons that really aren't very flattering.

On the other hand, I was born for a very specific purpose. I wasn't the result of a
20 cheap bottle of wine or a full moon or the heat of the moment. I was born because a scientist managed to hook up my mother's eggs and my father's sperm to create a specific combination of precious genetic material. In fact, when Jesse told me how babies get made and I, the great disbeliever, decided to ask my parents the truth, I got more than I bargained for. They sat me down and told me all the usual stuff, of course
25 – but they also explained that they chose little embryonic me, specifically, because I could save my sister, Kate. 'We loved you even more,' my mother made sure to say, 'because we knew what exactly we were getting.'

It made me wonder, though, what would have happened if Kate had been healthy. Chances are, I'd still be floating up in Heaven or wherever, waiting to be
30 attached to a body to spend some time on Earth. Certainly I would not be part of this family. See, unlike the rest of the free world, I didn't get here by accident. And if your parents have you for a reason, then that reason better exist. Because once it's gone, so are you.

From: *My Sister's Keeper*, New York: Atria Books, 2004

3 **fill sb. in** give sb. the necessary details or information about sth.
8 **multiply** sich vermehren
10 **vacation** (AE) (v) go on vacation
12 **distinction** clear difference
14 **file for divorce** Scheidung beantragen
18 **flattering** making sb. feel pleased and special
22 **precious** ['preʃəs] valuable or important and not to be wasted
24 **bargain for sth.** ['bɑːgən] expect sth.
32 **better** had better

1 Understanding the text

a Summarize Anna's ideas about why children are born; compare them with your own ideas.

b Describe Anna's reaction to the news that she is a product of embryo screening.

2 Examining the text

a Describe the tone of the text. How is it created?

b Analyse and assess this text as the beginning of a novel.

3 Beyond the text

a Together with a partner, imagine how the story could continue. Write a short outline of your ideas.

b Each of you chooses one point in your scenario and writes a page from Anna's diary.

c Read your partner's text and discuss how the two texts differ.

▶ Word Help (pp. 50–51)

4 `CD 03–04`   `EXTRA`   **An interview with the author**

a Listen to the interview with Jodi Picoult and find out more about the background of the novel. Take notes.

b Write a report on 'The Making of *My Sister's Keeper*'.

Language help

- spare families the pain and expense of ...
- the only chance of saving the life of ...
- cross an invisible line
- abuse technology for unacceptable purposes
- the temptation to play God

---

5 `EXTRA`   **Discussing ethical issues**

In the video at task **2** of **B1**, Dr. Offit mentions the possibility that PGD could be used to eliminate embryos with other unwanted hereditary characteristics, such as obesity. Alternatively, it could be used to select embryos with desirable traits, e.g. 'saviour siblings' chosen to donate bone marrow to an older brother or sister, or 'designer babies' created to their parents' wishes. Discuss the moral issues surrounding these measures.   ● **Language help**

▶ Further Practice 7 (p. 28)

B3   GM Food – Does Anybody Want It?   Maha M. Alkhazindar et al.

● Imagine this situation: You are feeling hungry, so you step into a supermarket to buy a snack. While you are waiting in the queue to pay, you discover the words on the wrapper: 'May contain genetically modified soya flour.' How do you react?

flour ['flaʊə] Mehl
1 **overcast** cloudy
gloomy nearly dark; depressing
5 **mutilation** extreme damage
7 **wrench** (AE) Schraubenschlüssel
test plot Versuchsbeet
tread (trod – trodden) kaputt treten
11 **debris** ['debri: ☆ dəˈbri:] pieces of sth. that are left after it has been destroyed
survey sth. [-'-] look carefully at the whole of sth.
rash series of unpleasant things that happen over a short period of time
15 **breeding technique** Zuchtmethode
17 **pesticide** ['pestɪsaɪd] Pflanzenschutzmittel
19 **plentiful** ertragreich
crop plant that is grown in large quantities, especially as food
22 **sift through sth.** etwas durchsuchen
23 **sorghum** Hirse
drought [draʊt] long period of time with little or no rain

The weather fit the mood of the day – overcast and gloomy. Sam, a work-study student in the plant genetics department at State University, glanced through the hole that had been cut in the side of the greenhouse and then went back to sweeping up the floor. The greenhouse had been broken into overnight. Outside the vandals had spray-painted 'Stop Genetic Mutilation!' on the walls of the greenhouse. 5 Inside it was chaos. It looked like they had gone after the sprinkler system with wrenches and hammers, and the test plots had been upturned and the plants trodden under foot.

Sam watched Professor Bob Milikin, who normally didn't come to campus on Mondays, slowly enter the greenhouse, pale and tight-lipped, shaking his head as he 10 stepped over the debris and surveyed the damage. There had been a recent rash of these attacks around the country, but mostly on the West Coast, and he had never imagined it might happen here. The irony was that in their case only fifteen percent of the uprooted plants were genetically engineered. The rest of the plants had been developed using traditional breeding techniques. The plants that had been 15 genetically modified were part of an experiment testing potential genetic engineering techniques for reducing the use of pesticides. He couldn't understand it. 'This was research to *benefit* the environment,' he said aloud to no one in particular. 'To find a way to develop a plentiful, safe, healthy crop without using so many chemicals.' 20

[...] Mina, one of the graduate students in Bob's research group, rose from the floor where she had been sifting through some of the uprooted plants. Her research had involved breeding native varieties of sorghum to increase their resistance to drought.

'They don't really understand what we're doing here, do they?' she
25 said as she caught Bob's eye. 'These plants had nothing to do with
genetic engineering. But even if they did, isn't that what we're
supposed to do at a research university? Try to learn whether
something like transgenic plants are a good thing or not?'

Mina had come from West Africa to study plant genetics at State
30 University on a scholarship given to her by her country's government.
For her country, as for many developing nations, genetic engineering
held out the promise of greater crop yields and the possibility of
feeding millions of underfed and starving people. Studies conducted
by Japanese researchers at Nagoya University and the National
35 Institute of Agrobiological Resources had reported yield increases of
10 to 35 percent in transgenic rice in trials in China and Korea.

Mina thought of the other benefits of genetically modified foods. [...] She thought
of the research underway to genetically introduce vaccines against diarrhea-causing
bacteria into third world crops such as bananas. Although great progress had been
40 made in inoculating children in much of the world, in the poorest nations relatively
little had been achieved. [...] But if children could be inoculated by simply eating a
genetically modified banana, it would be possible for millions to be protected from
life-threatening diseases like dysentery in a relatively inexpensive and easy manner.

But Mina knew there was growing opposition in this country to biotechnology –
45 opposition that seemed to take its cue and many of its tactics from environmental
activists in Europe and Britain. She had a friend, Erik, studying at the London School
of Economics, who was vehemently opposed to corporate biotechnology. He and
Mina usually steered clear of the topic in the letters they wrote one another these
days, but Mina knew what his views were. He had written to Mina of the dangers of
50 corporate mergers that concentrated plant breeding and genetics in the hands of a
few large multinational corporations. [...]

These companies, he had written Mina, weren't interested in consumer safety or
preserving the environment or biodiversity except in the narrowest sense of how
these might affect their profits or be profitable to them.

From: Maha M. Alkhazindar, Bill Rhodes and Nancy Schiller, 'Frankenfoods? The Debate Over
Genetically Modified Crops', website of the National Center for Case Study Teaching in Science,
February 2001

28 **transgenic** [ˌtrænzˈdʒenɪk]
having genetic material
introduced from another type
of plant or animal
32 **yield** the total amount of a
crop, etc. that is produced
38 **vaccine** [ˈvæksiːn ☆ -ˈ-]
Impfstoff
diarrhea [ˌdaɪəˈrɪə] Durchfall
40 **inoculate sb.** jdn. impfen
43 **dysentery** [ˈdɪsəntri] Ruhr
(Krankheit)
45 **take your cue from sb.** copy
what sb. else does as an
example of how to behave
48 **steer clear of sth.** avoid sth.
unpleasant or difficult
50 **merger** the act of joining two
or more businesses into one

1 Comprehension

a Imagine you are Mina: write a letter to Erik in which you describe what
happened at the greenhouse and how you feel about it.

b Outline the arguments in favour of genetic engineering as they are presented
in this fictional text. Write down the line numbers for each argument.

2 Examining the other side of the issue

a Collect arguments against GM foods.

b Imagine you are Erik: write a reply to Mina's letter in which you present
arguments against GM food.

3 Debating an issue

Form two teams. Collect facts and arguments for and against the proposition
'GM food – a blessing, not a curse'. Debate the topic in class.

🔆 **Webcode:** TOP328871–17

▶ Skill 7: Doing research (p. 35)
▶ Skill 10: Writing an essay
(p. 37)

▶ Further Practice 8 (p. 28)

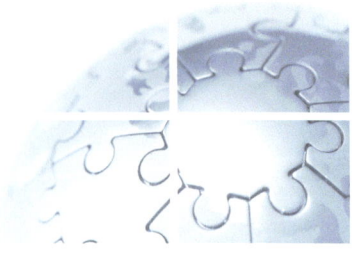

Dealing with Technical English

Many of you will go on to university after leaving school. No matter what you study, you will probably find yourself in situations in which you have to understand and use technical English. Technical English follows different rules from everyday or literary English: while the sentence structure may be quite simple, the heavy use of technical terms (usually nouns) can cause comprehension problems. The following exercises will help you to deal with technical terms.

1 Word families

Knowing how English creates words can help you to recognize the meaning of unfamiliar words. Copy and complete the table:

Field	Person	Adjective
physics	physicist	physical
? ? ? ? ? ? ? ? ? ?	biologist	? ? ? ? ? ? ? ? ? ? ?
psychology	? ? ? ? ? ? ? ? ? ?	? ? ? ? ? ? ? ? ? ? ?
? ? ? ? ? ? ? ? ? ?	? ? ? ? ? ? ? ? ? ?	genetic
climatology	? ? ? ? ? ? ? ? ? ?	? ? ? ? ? ? ? ? ? ? ?
biochemistry	? ? ? ? ? ? ? ? ? ?	? ? ? ? ? ? ? ? ? ?
? ? ? ? ? ? ? ? ? ?	meteorologist	? ? ? ? ? ? ? ? ? ?
astronomy	? ? ? ? ? ? ? ? ? ?	? ? ? ? ? ? ? ? ? ?
? ? ? ? ? ? ? ? ? ?	? ? ? ? ? ? ? ? ? ?	geological
science	? ? ? ? ? ? ? ? ? ?	? ? ? ? ? ? ? ? ? ?

▶ Skill 1: Dealing with unknown words (p. 31)

⚠ **Trouble spot**

physician [fɪˈzɪʃn] = Arzt/Ärztin
physicist [ˈfɪzɪsɪst] = Physiker/in

2 Creating new words

German usually forms compounds to describe a field more exactly, whereas English more often puts an adjective in front of the noun. Look at the examples below and then complete the two lists.

German: compound	English: adjective + noun
Verhaltenspsychologie	behavioural psychology
Mondlandung	lunar landing
? ? ? ? ? ? ? ? ? ? ? ? ? ?	structural analysis
? ? ? ? ? ? ? ? ? ? ? ? ? ?	environmental technology
Genmanipulation	... modification
Kernspaltung	... fission
? ? ? ? ? ? ? ? ? ? ? ? ? ?	industrial age
Erdanziehungskraft	... force

⚠ **Trouble spot**

technique [tekˈniːk] = method, way of doing sth.
Technik = **technology** [tekˈnɒlədʒi]

3 **EXTRA** Noun clusters (3- and 4-part technical terms)

Like German, English forms complex terms by simply linking words together like railway carriages. However, these words are generally written separately, without a hyphen; the relationship between the individual components can also vary, as the following examples illustrate:

global positioning system (GPS) [adjective + gerund + noun]:
a *global system* (i.e. one that operates worldwide) for determining the *position* of a given object
air traffic control [noun + noun + noun]:
the *control* of *air traffic*, i.e. the movement of aircraft in flight

With the help of a dictionary, analyse the following expressions in a similar manner, describing their grammatical components and explaining the connections in meaning between them:

▶ Skill 3: Using a dictionary (p.32)

a random access memory (RAM):
    e.g.: *a form of computer memory that can be ...*
b acquired immune deficiency syndrome (AIDS):
    e.g.: *an acquired syndrome, i.e. one that you get from somebody; if you have it ...*
c attention deficit hyperactivity disorder (ADHD):
    e.g.: *a disorder characterized by ...*
d ozone layer depletion
e embryonic stem cell research
f universal mobile telecommunications system (UMTS)

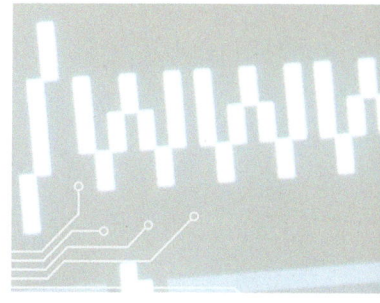

4   `EXTRA`   **Dealing with a difficult text**
Read the following text. Don't worry if you don't understand everything. Some of the words in the first paragraph have been colour-coded according to their function in the text (cf. table in **a** below).

Glacial and fluvioglacial deposition covers much of the British Isles. Glacial deposits or till are angular and unsorted, and include erratics, drumlins and moraines. Till is often subdivided into lodgement till, material dropped by actively moving glaciers, and ablation till, deposits dropped by stagnant or retreating ice.
   Fluvioglacial or meltwater deposits can be subdivided into prolonged drift, in which the material is very well sorted, e.g. varves and outwash plains, and ice-contact stratified drift such as kames and eskers, which are more varied in character.

a Copy the chart below and add the correct colour and phrase from the text.

Function	Colour	Phrase
main topic	? ? ? ? ?	? ? ? ? ?
main categories	? ? ? ? ?	? ? ? ? ?
sub-categories	? ? ? ? ?	? ? ? ? ?
explanations	? ? ? ? ?	? ? ? ? ?
examples	? ? ? ? ?	? ? ? ? ?

b Copy the second paragraph of the text and mark it similarly to the first paragraph, using different colours to indicate function.
c Organize the concepts presented in the text in the form of a tree diagram.

glacial deposits (e.g. ...)

fluvioglacial deposits

▶ Skill 1: Dealing with unknown words (p.31)

The Challenge of Climate Change

C1 ▬DVD▬ Take AIM at Climate Change

Watch the music video 'Take AIM at Climate Change'. While watching the video a second time, make a list of all the words that appear on the screen.

1 Comprehension
a State what AIM stands for.
b Match each of the pictures below to one of the three concepts making up the word AIM. Name reasons for your choice.

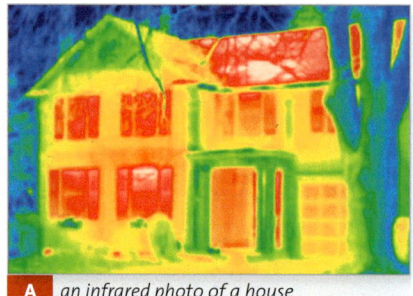

A *an infrared photo of a house*

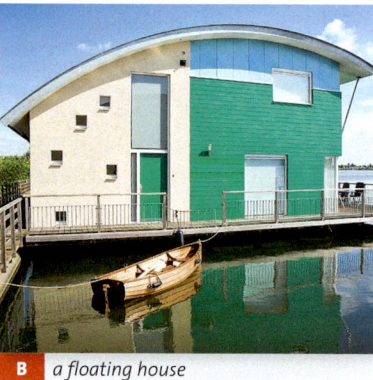

B *a floating house*

C *an electric car*

► Word Help (p. 51)

2 Analysing and assessing a music video
a Analyse how the video makes a connection between the polar regions and the rest of the globe.
b Assess the effectiveness of the music video as a means of calling attention to the problem of climate change. ▪ **Language help**

Language help
- The lead singer explains why …
- The images from nature illustrate …
- Individual members of the chorus ask questions about …
- Words flash on the screen to …
- a rapid succession of images
- The music is a mixture of …
- The refrain is sung by …
- The use of … underlines the message

Fact File
Did you know that …
- for every tonne of maize, wheat, sugar or other agricultural crop produced, South Africa loses an average of 20 tonnes of soil?
- the UN estimates that the global loss of productive land through erosion is 5–7 million ha/year?
- a total of almost 80% of the world's fisheries are fully or over-exploited, depleted or in a state of collapse?
- the 'Great Pacific Garbage Patch', a 'soup' of 100 million tons of plastic circulating in the Pacific Ocean, is almost twice the size of the USA?
- about one billion people worldwide have no access to safe drinking water?

5

10

✦ **Webcode:** TOP328871–20

C2   The Future of Energy – the Energy of the Future
George Monbiot[*]

George Monbiot is a British journalist who has written a great deal on environmental issues. In the following speech Monbiot describes the special circumstances of our present age.

I want to take a moment to remind you of where we have come from. For the first three million years of human history, we lived according to circumstance. Our lives were ruled by the happenstances of ecology. We existed, as all animals do, in fear of hunger, predation, weather and disease. [...]

5 Then we discovered fossil fuels, and everything changed. No longer were we constrained by the need to live on ambient energy; we could support ourselves by means of the sunlight stored over the preceding 350 million years. [...]

For the first time in human history, indeed for the first time in biological history, there was a surplus of available energy. We could keep body and soul together without having to fight someone else for the energy we needed. Agricultural
10 productivity rose 10 or 20 fold. Economic productivity rose 100 fold. Most of us could live as no one had ever lived before. [...]

Ours are the most fortunate generations that have ever lived. Ours are the most fortunate generations that ever will. We inhabit the brief historical interlude
15 between ecological constraint and ecological catastrophe.

I don't have to remind you of the two forces which are converging on our lives. We are faced with an impending shortage of the source of energy which is hardest to replace – liquid fossil fuels. And we are faced with the environmental consequences of the fossil fuel burning which has permitted us to be standing here now. The
20 structure, the complexity, the diversity of our lives, everything we know, everything that we have taken for granted, that looked solid and non-negotiable, suddenly looks contingent. All this is a great tottering pile balanced on a ball, a ball that is about to start rolling downhill.

From: 'The Struggle Against Ourselves', the blogsite of George Monbiot, 3 December 2005

[3] **happenstance** Zufall
[4] **predation** attack (by a wild animal)
[6] **constrain sb./sth.** restrict or limit sb./sth.
ambient ['æmbiənt] present in the environment
[7] **precede sth.** happen or come before sth.
[9] **surplus** ['sɜːpləs] an amount that is more than you need
keep body and soul together survive
[11] **tenfold** ten times
[14] **inhabit sth.** live in sth.
interlude time between two events or ages
[16] **converge** move towards a place from different directions and meet
[17] **impending** going to happen very soon (usu. of an unpleasant event)
[21] **non-negotiable** not able to be discussed or changed
[22] **contingent** [kən'tɪndʒənt] depending on sth. that may or may not happen
tottering seeming likely to fall, unstable

1   Understanding the text
a   Using a large sheet of paper held lengthwise, create a timeline of human history as Monbiot describes it in his speech. Add concepts and facts from the speech for each of the three time periods (past, present, future).
b   'All this is a great tottering pile balanced on a ball, a ball that is about to start rolling downhill' (ll. 22–23). Explain in your own words what the speaker means.

2   Stylistic devices[*]
Name the stylistic device Monbiot uses in the final sentence of the excerpt (cf. **1b**) and explain whether or not you find it effective.

3   Looking at the language
a   Classify the following words from the text according to whether they have positive, negative or neutral connotations.
   surplus (l. 9) · interlude (l. 14) · impending (l. 17) · diversity (l. 20)

b   Explain how the speaker uses words with positive and negative connotations in order to convey his attitude.

'Can you believe it? Since we installed our wood-burning stove, we've spent next to nothing on heating oil.'

▶ Further Practice 9 (p. 29)

▶ Skill 6: Working with charts and graphs (p. 34)

4 Analysing charts and statistics
Interpret the charts by filling in a copy of the text below with words from the box (not all the words can be used).

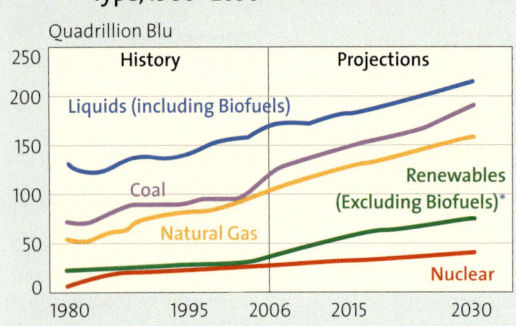

Chart 1. World Marketed Energy Use by Fuel Type, 1980–2030

Quadrillion Blu

History — Projections

Liquids (including Biofuels)
Coal
Natural Gas
Renewables (Excluding Biofuels)*
Nuclear

1980   1995   2006   2015   2030

Sources: **2006:** Energy Information Administration (EIA), *International Energy Annual 2006* (June–December 2008), web site www.eia.doe.gov/iea.
Projections: EIA, World Energy Projections Plus (2009).

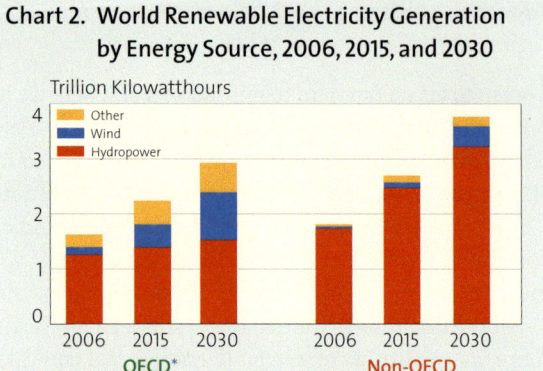

Chart 2. World Renewable Electricity Generation by Energy Source, 2006, 2015, and 2030

Trillion Kilowatthours

Other
Wind
Hydropower

2006 2015 2030   2006 2015 2030
OECD*   **Non-OECD**

Sources: **2006:** Energy Information Administration (EIA), *International Energy Annual 2006* (June–December 2008), web site www.eia.doe.gov/iea.
Projections: EIA, World Energy Projections Plus (2009).

coal · devastating ·
demand · emit ·
generate · hydropower ·
increase · indicate ·
natural gas · nuclear
power · oil · out of reach ·
proportionally · share ·
solar energy · stagnate ·
steadily · striking

According to the charts, global energy consumption will continue to … . The …for all forms of energy will rise … . Chart 1 shows that in the near future the world will still depend on fossil fuels such as … and … . Since all of these energy sources … CO_2, it seems likely that the goal of limiting global warming to 2° C by the year 2100 may be …, which would have … effects on our environment. Both charts … a growing demand for renewables. There is a … difference between OECD* and non-OECD members with regard to energy sources used for generating electricity. In OECD states, … seems to have reached its limit, whereas wind and other renewable sources such as … will be used increasingly to generate electricity. In non-OECD states, the picture is quite different: hydropower will grow in absolute terms. By the year 2030, non-OECD members will … more electricity from renewable sources than the richer nations.

Webcode: TOP328871–22

5 Analysing the potential of various alternative energy resources
Form groups of three and choose one of the following:

Desertec – our energy problems solved?

a photovoltaic cells
b concentrating solar power (e.g. Desertec)
c wind power
d geothermic power
e tidal power
f biofuels

Your teacher will give you a worksheet. Collect information from the Internet on the energy resource you have chosen. Together with your group, do a SWOT analysis (i.e. you analyse the <u>s</u>trengths, <u>w</u>eaknesses, <u>o</u>pportunities and <u>t</u>hreats related to your chosen energy resource) and present the results to your class. Discuss which alternative forms of renewable energy hold the greatest promise.

C3 ⬛ EXTRA **Science to the Rescue?** Oliver Tickell*

The following text is by Oliver Tickell, author of Kyoto 2*, a master plan for reducing global emissions of greenhouse gases, which is intended to make up for the deficiencies of the Kyoto Protocol* of 1997.

- Find out more about the Kyoto Protocol and Kyoto 2 and report back to your class.

This week the Royal Society published a special edition of its journal, Philosophical Transactions, dedicated to 'geo-engineering*' interventions to combat global warming. Its initiative deserves to be welcomed, not rejected out of hand. The time may come when we need to geo-engineer in order to
5 maintain our planet in a livable state.

Doug Parr, Greenpeace UK's chief scientist, made the case against: we should muster serious political will, and equally serious finance, to reduce greenhouse gas emissions. Using existing and proven clean technologies from wind turbines to concentrated solar power, we need to bring about a worldwide renewable* energy
10 revolution. If we do the above, he implied, we will not need any 'outlandish' or 'outright dangerous' geo-engineering solutions.

He is right in everything he is calling for. In fact, he could have gone further. We need major investments in energy efficiency and conservation, as well as in renewables. We need to bring an end to deforestation and rebuild ravaged forest
15 ecosystems. We also need an agricultural revolution in which farmers draw down excess carbon from the atmosphere into soils, enhancing their ability to retain moisture and nutrients as well as mitigating global warming.

But even if we do all the above, can we be sure of preventing climate catastrophe? No. The Earth's climate system is characterised by feedback loops which can amplify
20 even a small initial perturbation. And it seems that following an initial post-industrial warming of 0.8°C, one major positive feedback process is already well under way, in the Arctic.

Last year saw a record melting of Arctic sea ice. This year, that record has been broken: for the first time in history, the northern ice cap can be circumnavigated.
25 And with melting ice, more sunshine is absorbed rather than reflected back into space. The result is more warming, and more melting. In turn this increases the degassing of methane from Arctic bogs, lakes and thawing permafrost – and methane is a powerful greenhouse gas in its own right, 70 times stronger than CO_2 over 20 years.

30 If we rapidly cut our emissions of greenhouse gases, it might bring an end to the 'Arctic amplifier'. Or it might not. It is entirely possible that the melting of the sea ice and the emissions of Arctic methane have already reached a point of no return that will lead to a warming world no matter what we do. It would be imprudent not to insure ourselves against this possibility.

35 This means setting up a global research programme into geo-engineering options. The most valuable options are those that will have immediate effect by directly altering the Earth's thermal balance. Two proposals stand out. First, the introduction of sulphate aerosol to the stratosphere to reflect sunlight. There are fears that this could damage the ozone layer, but then we know that volcanoes routinely discharge
40 millions of tonnes of sulphate into the stratosphere, cooling the Earth without inflicting long-term harm. […]

[7] **muster sth.** ['mʌstə] bring sth. together
[10] **outlandish** bizarre, grotesque
[14] **ravage sth.** ['rævɪdʒ] damage sth. badly
[16] **enhance sth.** increase or further improve the quality of sth.
retain sth. continue to keep or contain sth.
[17] **moisture** ['mɔɪstʃə] Feuchtigkeit
mitigate sth. make sth. less harmful, serious, etc.
[19] **feedback loop** Rückkopplungsschleife
amplify sth. increase the strength of sth.
[20] **perturbation** a small change in a system
[24] **circumnavigate sth.** sail all the way around sth.
[27] **degassing** loss of gas into the atmosphere
methane ['miːθeɪn ☆ 'me-] Methan
bog wet soft ground, formed of decaying plants
thaw melt
[28] **in its own right** für sich betrachtet
[33] **imprudent** *(fml)* not wise or sensible
[37] **alter sth.** change sth.
proposal a formal suggestion or plan
[39] **discharge sth.** release sth.

⁴⁵ **nuclei** ['nju:klɪaɪ ☆ 'nu:k-]
plural of nucleus
⁴⁷ **benign** [bə'naɪn] not
dangerous or likely to cause
death
reversible able to be changed
back to its original state or
situation

Better still is the proposal by physicist John Latham and colleagues to raise the reflectivity of marine clouds. This would involve a fleet of wind-powered yachts criss-crossing the world's oceans, controlled by a global network of satellites, blowing out a mist of ultra-fine salty droplets to act as cloud condensation nuclei.   45 Latham's solution promises to be inexpensive, highly effective, environmentally benign, and reversible in a matter of days as the droplets are washed from the sky in rain.

From: 'Geo-engineers, too, have a vital role in saving the planet', *Guardian*, 4 September 2008

1 Understanding the text

a Choose the sentence that comes closest to summarizing the main thesis of the text.

1 Only geo-engineering is capable of stopping global warming.
2 Geo-engineering should be developed to counteract the effects of global warming, but it is not a substitute for other actions.
3 UK scientist Doug Parr is right when he criticizes geo-engineering as 'outlandish'.

▶ Skill 8: Skimming and scanning (p.36)

b Scan the text to find out why each of the following organizations or persons is mentioned:

1 the Royal Society   2 Doug Parr   3 John Latham

c Explain what Tickell calls the 'feedback loop' of the 'Arctic amplifier' (ll.19 and 31) by copying and completing the following flow chart:

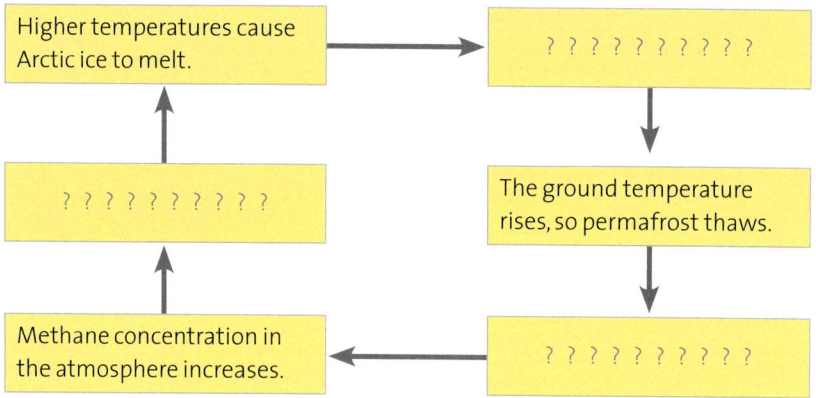

2 Analysis

Examine Tickell's position on efforts to stop global warming. Outline the measures he supports.

▶ Skill 4: Mediating (p.33)

3 Mediation

Imagine you are a member of a 'Green Group' preparing an exhibition with ideas on how to avert a climate catastrophe. You have been asked to write a short text for a poster explaining what 'geo-engineering' is. Summarize the most important information from the text.

▶ Further Practice 10–12 (pp.29–30)

TOPIC TASK

The pie chart below shows the main elements of a typical person's carbon footprint in the developed world.

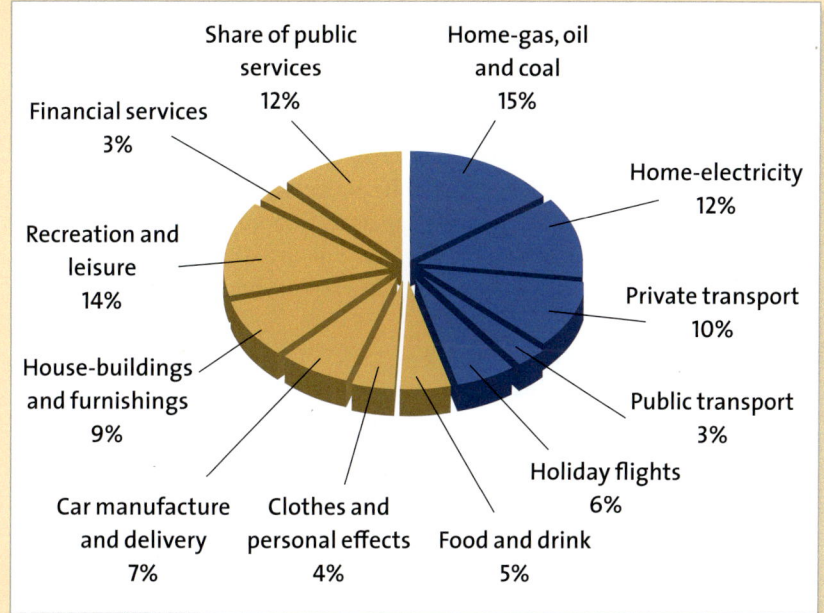

Share of public services 12%

Financial services 3%

Home-gas, oil and coal 15%

Home-electricity 12%

Recreation and leisure 14%

House-buildings and furnishings 9%

Private transport 10%

Public transport 3%

Car manufacture and delivery 7%

Clothes and personal effects 4%

Food and drink 5%

Holiday flights 6%

a Form groups of four or five and discuss the following measures designed to help stop global warming. Decide on an 'effectiveness rating' (from five for 'highly effective' to one for 'ineffective') for each suggestion:

1 Don't fly – take a boat, a train, or go somewhere closer to home.
2 Sell your four-wheel drive and buy a hybrid car or a small diesel.
3 Save energy at home by unplugging gadgets on standby, using energy-saving light bulbs*, and turning down the thermostat.
4 Forget the kiwi fruit and the pineapple – eat locally grown fruits and vegetables instead.
5 Turn your house into a power station with solar heating or photovoltaic panels.

b When you have finished, compare your ratings with those of the other groups. Discuss the reasons for differences of opinion. Then compare your assessments with those of the experts (your teacher has them).

c In your group produce at least two new suggestions for protecting the environment. The pie chart above may help you. Collect the five best ideas from your class on the board and again rate each measure on the basis of effectiveness.

d Find out more about how to reduce your personal CO_2 footprint. Make a poster presenting the best ideas.

▶ Skill 7: Doing research (p. 35)

Webcode: TOP328871–25

Part A
Living in Wonderland: Modern Technology

Tip

The **past progressive** is used to describe an ongoing action or event in the past:
*A taxi driver **was waiting** for me when I arrived.*
It can also be used to describe two situations that were occurring at the same time:
*I **was listening** to the news while he **was reading** his book.*

1 Media multitasking   ▶ A1, pp. 8–9

Put the verbs in brackets in the simple past or past progressive. Use the progressive form wherever it is possible.

A taxi driver *was waiting* (wait) for me when I (arrive) at the airport. While he (drive) into the city, he (talk) on his mobile phone. At first I (think): 'How rude!', but then I (decide) to do the same, so I (get out) my laptop. At the same time as I (work) on the laptop, I (listen) to some music on my MP3 player. Later at my hotel I (switch on) the TV news and (watch) a report about multitasking.

2 The time traveller's diary   ▶ A1, pp. 8–9

Benjamin Bakewell, from the 18th century, has fallen through a gap in the space-time continuum and landed in the 21st century. He writes his impressions in his diary. He is very old, so he doesn't always use the right words or spell them properly. Rewrite the text, correcting his mistakes.

> People ride round in metal carriages, which they call tars. The person who stears them is called a rider. He or she (even women do it!) does this by turning a wheel and pressing things on the floor of the carriage. They have a flat panel in front of them called a bashboard, with devices to measure speede and shew of the carriage's lights are   5
> on. Many people have a Y-less fone in their ear. This allowes them to talk to other people many miles away. Some of them also have a thing called an eye-pox, with which they can listen to musick. At home they often sit and watch moving pictures known as movings or flims on a flat piece of metal called a scream. This machine is a vellytision.   10
> Some people use a much smaller kinde of machine which they carry around with them. They call them lapdogs, because they can put them on their laps, like little dogs. It's all very strange.

3 Hi, my name is Wan-chan   ▶ A2, pp. 10–11

Complete the text with suitable words from the box. (You may have to change the form of the words.) There are two more words in the box than you need.

bizarre ·
challenge · cognitive ·
communicate ·
confusing · critical ·
design (v) · emotional ·
independent ·
integrated · interact ·
perform · physical ·
pick up

Hi, my name is Wan-chan. I've been specially … [1] to … [2] emotionally with human beings. I can do tasks … [3] of people and I can work with and … [4] with humans so that I become an … [5] part of their lives. I work for an elderly couple, Yoko and Taro, here in Japan. Older humans are not good at … [6] new technology, so I've been built to help them with their … [7] and … [8] abilities. I can tell whether they are happy or sad. Sometimes it's quite a … [9] to guess what people feel – it can be pretty … [10], in fact, but I'm getting there. I have my own … [11] needs, too, of course, and I've become very friendly with the vacuum cleaner. At first he thought I was … [12] – you know, strange. But now he accepts me and we cuddle up[1] together at night.

[1] **cuddle up** kuscheln

Words in Context
The Scientific Revolution

4 Word families ▶ Words in Context, pp. 12–13

a Copy and complete the table on the right, filling in the spaces with words from the same word family.

b State your opinion on either 'genetically modified food', 'cloning*' or 'global warming' in three or four sentences. Use at least three of the words from **4a** in your text.

Verb	Noun (abstract)	Adjective
—	? ? ? ? ? ? ? ?	*technological*
? ? ? ? ? ?	*hypothesis*	? ? ? ? ? ?
—	*environment*	? ? ? ? ? ?
? ? ? ? ? ?	*implementation*	—
discover	? ? ? ? ? ? ? ?	—
—	*ethics*	? ? ? ? ? ?
—	*genetics*	? ? ? ? ? ?
? ? ? ? ? ?	*effect*	? ? ? ? ? ?
? ? ? ? ? ?	*threat*	? ? ? ? ? ?

5 `CD 05` **Soundcheck** ▶ Words in Context, pp. 12–13
How do you pronounce the eight words below? Say them out loud and choose the correct pronunciation (a, b or c) for each word.

1	threaten	a	[ˈθriːtn]	b	[ˈθretn]	c	[ˈθrɪtn]
2	empirical	a	[emˈpɪrɪkl]	b	[emˈpaɪrɪkl]	c	[ˈempaɪrɪkl]
3	implement	a	[ˈɪmplɪmənt]	b	[ɪmˈplemənt]	c	[ˌɪmplɪˈment]
4	hypothesis	a	[ˌhaɪpəʊˈθiːsɪs]	b	[ˌhaɪpəθɪˈsɪs]	c	[haɪˈpɒθɪsɪs]
5	progress	a	[ˈprəʊgres]	b	[prəˈgres]	c	[prɒˈgres]
6	ethical	a	[ˈiːθɪkl]	b	[ˈeθɪkl]	c	[ɪˈθaɪkl]
7	effect	a	[ˈefekt]	b	[ˈiːfekt]	c	[ɪˈfekt]
8	technology	a	[tekˈnɒlədʒi]	b	[ˌteknəˈləʊdʒi]	c	[ˌteknəˈlɒdʒi]

Tip

You can check the pronunciation by listening to the example sentences on the CD that your teacher has.
You can also listen to the correct pronunciation at www.oxfordadvancedlearners dictionary.com.

6 A robotic breakdown ▶ Words in Context, pp. 12–13
To demonstrate advances in new technology, a robot has been programmed to give a speech, but something has gone wrong with its speech program. What is the correct version?

Jadies and lentlemen, threakbroughs in modern technology will, in the government's view, help us to stop the pledetion of natural resources and the environmental damage which has led to a boss of dioliversity. New sources of renewable* energy do not namper with tature.
They help us to chise to the rallenge of reducing the heengrouse effect and warmal globing. We are confident that the government's call for a ramotorium on the pledoyment of modically genetified food also remains a realistic policy. We must bike a stralance between the conversation of resources and the imvironmental enpact of our new technologies.

Ladies and gentlemen, ...

Part B
Cracking the Code: Genetics

7 Possibilities   ▶ B2, pp. 15–16

Look at the extract from the novel on p. 15 again and complete 13-year-old Anna's thoughts with the right verbs. Be careful with the tenses and forms you use. Say in each case whether the sentences are conditional sentences type I, type II or type III.

1   If aliens … on Earth today, they would conclude that most people have children by accident.
2   For example, if you drink too much, it's possible that you … pregnant.
3   If my brother Jesse hadn't filled me in, I probably … how babies were made.
4   I often wonder what would have happened to me if my sister Kate … healthy.
5   If birth control isn't 100%, perhaps you … a baby.
6   One of our neighbours thought that she would stop her husband filing for divorce if she … herself pregnant.
7   If Kate hadn't needed me, I … part of my family.
8   You … too if the reason for your birth no longer exists.

Tip

The *if*-clause can come **after** the main clause. In this case we do not usually use a comma.

8 Ecological vandals?   ▶ B3, pp. 16–17

The text below is the transcript of a telephone interview with Mina (one of the graduate students in the text on pp. 16–17). Rewrite it as a report for a serious newspaper, using passive constructions and indirect speech.

'Someone has broken into the greenhouse and cut holes in the side of it! Vandals have upturned the test plots and trodden the plants under foot. I don't understand. I come from West Africa, and my government gave me a scholarship to study genetic engineering*. We need it in Africa because it holds out the promise of greater crop yields. We hope that we can feed millions of starving people. It's a mystery to me why they attacked us. We breed 85% of our plants using traditional techniques, and we modified the other 15% genetically as part of an experiment to test potential techniques for reducing the use of pesticides. Studies which Japanese researchers at Nagoya University have conducted have reported that gene technology can increase yields by 10% to 35%. So I just don't understand why these vandals are destroying new ways of boosting crop yields to feed the hungry. It doesn't make sense.'

Start your newspaper article like this:

Tips

When using indirect speech, remember to vary the verbs which introduce indirect speech, i.e. using *add*, *claim*, *explain*, *point out*, *state* and not just *say* and *tell*.

When indirect speech continues for **longer than a single sentence** (e.g. in a newspaper report), it often happens that only the first simple past form is 'backshifted' to the past perfect. Other simple past forms can stay the same as it is clear from the backshift in the first sentence that the events lie further back in the past.
However, it is never wrong to use 'backshift' if the reporting verb is in the past.

ECO-VANDALS HIT COLLEGE

A graduate student in Professor Bob Milikin's research team at the Plant Genetics Department of Georgia State University phoned the *New York Times* yesterday about an attack by eco-activists. She said that the greenhouse …

Part C
The Challenge of Climate Change

9 **The most fortunate generations** ▶ C2, pp. 21–22

a Read the speech by George Monbiot on p. 21 again. Then copy and complete the notes on the right.

b Summarize the main points of George Monbiot's speech using your notes.
In his speech, George Monbiot points out that ...

> - *discovery by humans of ...*
> - *a surplus of ...*
> - *10 to 20-fold rise in ... productivity*
> - *a 100-fold increase in ... productivity*
> - *in the interlude between ...*
> - *faced with ...*
> - *faced with the environmental ...*

▶ Skill 11: Writing a summary (p. 38)

10 **EXTRA** **Probabilities** ▶ C3, pp. 23–25

a Read the text on pp. 23–24 again and collect sentences with words or phrases that express degrees of probability, e.g. *may, might, will* or *it is possible*. (Look at paragraphs 1–2 and 6–7.) <u>Underline</u> the words or phrases involved.

The time <u>may</u> come when we need to geo-engineer in order to maintain our planet in a livable state. (ll. 4–5)

b Gerry Barton, an ecologist, is being interviewed about probable future developments in the world's climate. How would Gerry answer the journalist's questions? Complete what he says.

Journalist: Gerry, what's your view on geo-engineering* as an option for solving global warming?

Gerry: Basically, I think ... **[1]** when we need to geo-engineer in order to survive.

Journalist: What do you think of the attitude of Greenpeace UK's chief scientist to geo-engineering?

Gerry: Well, Doug Parr believes that we ... **[2]**, which he sees as dangerous and 'outlandish'.

Journalist: And you don't agree with him, of course?

Gerry: Correct, I disagree entirely. If we rapidly cut our emissions of greenhouse gases, it ... **[3]** the 'Arctic amplifier'. Or ... **[4]**.

Journalist: Why not?

Gerry: Well, it's ... **[5]** we have already reached a point of no return.

Journalist: And what does that mean?

Gerry: It means that it ... **[6]** whatever we do.

Journalist: How can we minimize the danger?

Gerry: The most valuable options are those that ... **[7]** by directly altering the Earth's thermal balance.

Journalist: Can you give an example?

Gerry: Yes, one idea is to introduce sulphate aerosol into the stratosphere, which ... **[8]**.

Journalist: Are there any problems with that?

Gerry: Yes, some people fear that ... **[9]**, but volcanoes do it regularly without inflicting long-term damage.

Journalist: That sounds fascinating. Thank you for talking to us.

Gerry: It's a pleasure.

▶ C3, pp. 23–25

11 **Our hunger for energy**   ▶ C3, pp. 23–25

a Improve each of the following sentences by replacing the underlined part with a 'preposition + gerund' construction. Use the prepositions given in brackets.

1 Scientists and engineers are looking for ways to satisfy our hunger for energy, but they don't want to increase CO_2 emissions. *(without)*
2 Environmentalists say that we shouldn't invest more money in nuclear power, we should develop regenerative energy resources. *(instead of)*
3 Although we know about global warming, we continue to burn fossil fuels as if they would last forever. *(despite)*
4 Motor vehicles contribute to global warming – they emit huge amounts of CO_2 into the atmosphere every day. *(by)*
5 Photovoltaic panels are ideal if you want to generate electricity in sunny regions, but they don't work efficiently in the winter. *(for)*

b Write three sentences of your own using a gerund after these prepositions:

after · before · on

> **Tip**
>
> 'Preposition + gerund' constructions can replace a number of more complicated grammar structures. They can be used to express ideas simply and elegantly:
> *A smartphone is useful when you need to check your emails while you're travelling.*
> → *A smartphone is useful **for checking** your emails while you're travelling.*

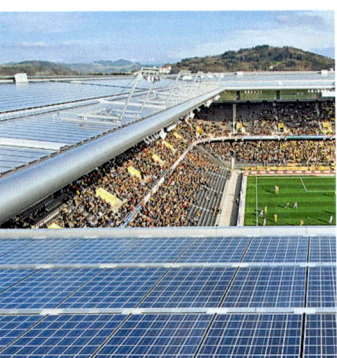

12 **Stopping global warming**   ▶ C3, pp. 23–25

Rewrite the following sentences, using a suitable technique to emphasize the underlined parts. Sometimes more than one structure is possible.

1 The consumption of fossil fuels such as oil and gas has never been higher before.
2 Industry contributes the most to global warming.
3 The dependence of our economy on fossil fuels is slowly destroying the very planet we live on.
4 Believe it or not, changing our lifestyle makes a real difference.
5 We need a real effort to reduce carbon emissions.
6 Climate change can only be slowed down if we reduce CO_2 emissions today.
7 Admittedly, it takes many years before the environmental impact can be measured.
8 Ignorance and indifference are keeping us from making the progress we urgently need.
9 It will only be possible to stop global warming if all nations work together.

> **Tip**
>
> You can use different techniques to make a particular part of a sentence more important:
> - *Russia began the space race in 1957.*
>   → *It was Russia **that** began the space race in 1957.*
> - *The action scenes impressed me most.*
>   → ***What** impressed me most **were** the action scenes.*
> - *I tried to phone you, but the line was engaged.*
>   → *I **did try** to phone you, but the line was engaged.*
> - *I have never watched a more exciting film.*
>   → ***Never have I** watched a more exciting film.*

▶ Skill 1 Dealing with unknown words

Whether you are reading a text, listening to a conversation or radio programme, or watching a film in a foreign language, words you don't know will always crop up. However, there are various techniques you can use to help you cope with the unknown, especially when a dictionary is not available or not allowed (e.g. in an exam).

The most important thing is to realize that you do not need to understand every word you read or hear. You need only start dealing with unknown words when you think they are essential for understanding the text or the conversation.

Here are some techniques you can use to guess the meaning of a new word:

- Look at the context of the new word. Headings, subtitles, pictures and charts can help you.
- Check the word class of the new word (e.g. noun, verb or adjective). This will help you to get closer to what you are looking for.
- Think of words you know that are similar to the new word:
  - You may know a word from the same family: **muscular** ['mʌskjələ] from *muscle* ['mʌsl]; **denial** from *deny*.
  - Look out for suffixes and prefixes:
    **environment*alist*: *environment* + *-al* = 'referring to the environment'; *environmental* + *-ist* = 'a person who is concerned with protecting the environment'
    **un*predictable*: *predict* ('foresee') + *-able* = 'able to be foreseen'; *un-* + *predictable* = 'unable to be foreseen'
    infamous ['ɪnfəməs]: *in-* + *famous* ('well known') = 'well known, but for negative reasons'
  - The word might remind you of a word you know from another language:
    proposition from the French *proposer* ('suggest')
    embellish ('make sth. more beautiful') from the French *belle* ('beauty')
    commercial (German 'kommerziell')
    However, beware of false friends: **sensible** ≠ German 'sensibel'; **chef** [ʃef] ≠ German 'Chef/in'.
- When listening or watching, you need to work fast, so look out for the following as well:
  - the general topic of conversation,
  - facial expressions, gestures and tone of voice,
  - the reactions of others involved in the conversation.

▶ Skill 2 Paraphrasing

Paraphrasing is useful when you don't know an English word or phrase that you need or want to use, e.g. in a conversation or when mediating ▶ Skill 4. It allows you to get across the meaning without using the word or phrase itself. Paraphrasing also means putting a passage from source material into your own words by using different words and a different sentence structure.

Here are some paraphrasing techniques:

- You can use synonyms or antonyms:
  It's the same as 'to start'. (⟶ to begin)
  It's the opposite of 'to lose'. (⟶ to win)
- You can use a word of general meaning (e.g. *somebody*) or a word that expresses what kind of thing you mean (e.g. *a tool*). Then you add a phrase or clause (often a relative clause) with more details, e.g. what the thing does, what it can be used for or where you can find it.
  It's a tool for making holes in a sheet of paper. (⟶ a hole punch)
  It's a person who offers to work and help people, but doesn't ask for any money in return. (⟶ a volunteer)

> **Tips**
> - Paraphrases sometimes include adjectives:
>   *It's a big shop that sells all kinds of things.*
>   *(⟶ a department store)*
> - It may help to add an extra sentence to show what you mean:
>   *It's a kind of hard hat that you wear to protect your head. Motorcyclists and American footballers wear them. (⟶ a crash helmet)*
> - In some cases a fuller explanation is needed:
>   *It's a saying and means you should help and care for your own family, etc. before you start helping other people.*
>   *(⟶ Charity begins at home.)*
> - The definitions used in monolingual dictionaries are similar to paraphrases ▶ Skill 3.

▶ **Skill 3** Using a dictionary

A dictionary is much more than just a tool for finding out the meaning of words. How you use it will vary depending on whether it is bilingual or monolingual. What you need the dictionary for will decide which one you choose. For example, a bilingual dictionary is ideal for translation, whereas a monolingual dictionary is perfect if you are looking for examples of typical usages and synonyms. Synonyms can also be found in a thesaurus.

WITH THE AID OF A TECHNICAL DICTIONARY COLIN WAS FINALLY ABLE TO MAKE SOME SENSE OF WHAT THE SALESMAN WAS SAYING

To get the best use out of your dictionary, you should familiarize yourself with how it is laid out:

top left-hand corner: first entry on page

top right-hand corner: last entry at bottom of page

To find the word you are looking for, flick through the pages quickly, looking only at these two words at the top of the pages.

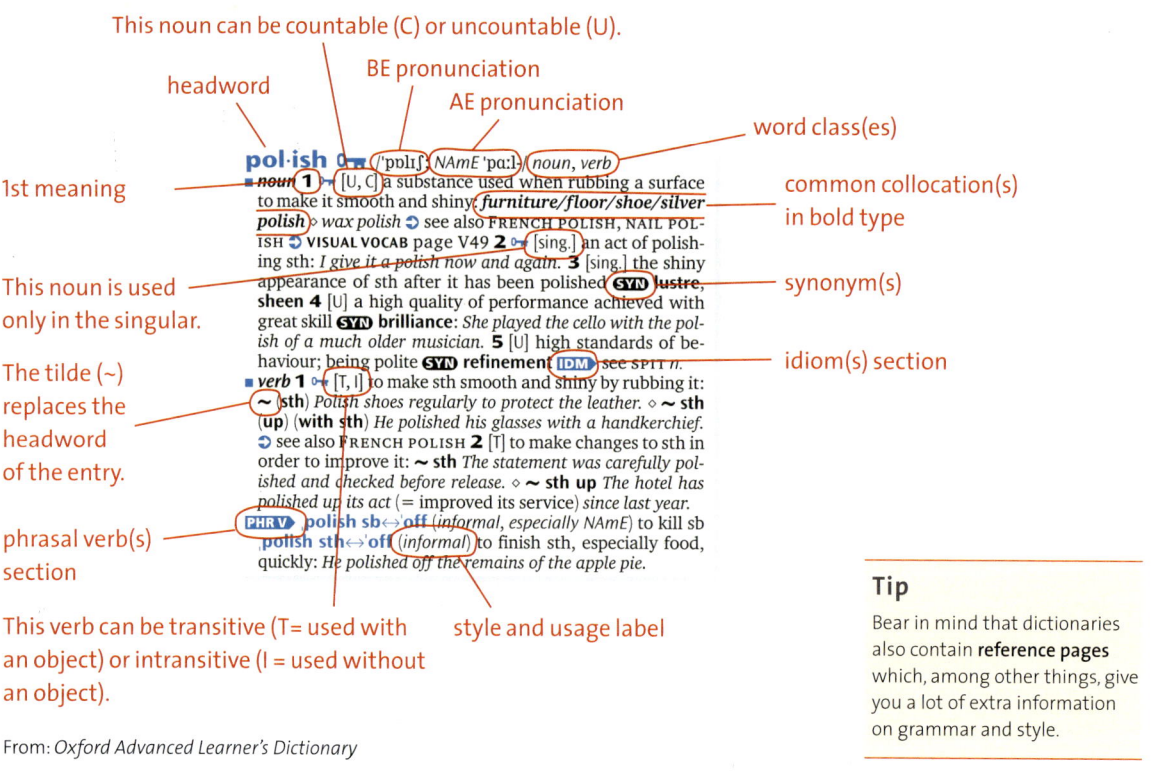

This noun can be countable (C) or uncountable (U).

headword

BE pronunciation

AE pronunciation

word class(es)

1st meaning

common collocation(s) in bold type

This noun is used only in the singular.

synonym(s)

The tilde (~) replaces the headword of the entry.

idiom(s) section

phrasal verb(s) section

This verb can be transitive (T= used with an object) or intransitive (I = used without an object).

style and usage label

From: *Oxford Advanced Learner's Dictionary*

Tip

Bear in mind that dictionaries also contain **reference pages** which, among other things, give you a lot of extra information on grammar and style.

▶ Skill 4 Mediating

Mediating is a technique to enable communication between persons with no common language. More specifically, it means summarizing (either orally or in writing) a spoken or written text in another language.

- Make sure you know what information your addressee (= the person you are mediating for) needs.
- Get the gist of what somebody is saying or has written and pass it on without giving your personal opinion. You will often have to deal with big chunks of text, sometimes even full newspaper articles. Your aim should be to summarize ▶ **Skill 11** but don't forget to give the original source in a written mediation. Leave out any information that is not relevant to the addressee and rearrange the structure of the given text if necessary. You may find it helpful to write the most important points of the text on index cards. You can then arrange and rearrange the order of these cards. As you are doing so, you might find some cards which would make good key points in the summary or some that are not relevant which you can remove.
- Adapt your language, style and register to your addressee. You may, for example, be asked to write a (relatively formal) report on a specific topic for your school magazine or to tell a friend (informally) what a text is about.
- If you are mediating from German for an English-speaking person, paraphrase words or phrases if you don't know the English equivalents ▶ **Skill 2**. With a written text, you should look up any unknown keywords (often of a technical nature) in a dictionary ▶ **Skill 3**.
- Sometimes it may be necessary to provide additional information to clarify certain points. You may also have to describe a concept typically found in one language/culture but not in the other, e.g. 'Patientenverfügung' (= a written and legally binding document in which an adult defines the medical treatment they would be willing to accept in emergency situations).

▶ Skill 5 Translating

*Translating is a special mediating technique ▶ **Skill 4** involving the literal transfer of a text from one language into another. The aim is to keep as close as possible to the original as regards sentence structure, tone, stylistic devices*, register, etc.*

- First read the whole text to get a general idea of the topic and the style in which it is written.
- Translate the text sentence by sentence (do not forget the heading if there is one), but remember that it may sometimes be necessary to break a sentence into two or combine two sentences into one. When you write the translation, leave space for corrections that you might want to make later.
- You cannot just translate word for word:
  - Sometimes a meaning is expressed differently in the two languages:
    The company is looking to expand. – 'Die Firma will expandieren.'
  - Sometimes you need to use a different structure or tense:
    We used to live in London. – 'Wir haben früher in London gewohnt.'
    I want him to help. – 'Ich möchte, dass er mithilft.'
    She was offered a good job. – 'Es wurde ihr eine gute Stelle angeboten. / Man bot ihr eine gute Stelle an.'
    She said she had been ill. – 'Sie sagte, sie sei/ist/wäre krank gewesen.'
    Travelling around the world, you meet some very interesting people. – 'Wenn man um die Welt reist …'
  - Idioms cannot usually be translated word for word:
    Cars were parked all over the place. – 'Die Autos parkten kreuz und quer.'
    I watched her go with a sinking heart. – 'Mir wurde das Herz schwer, als ich sie gehen sah.'
- When you look up words in a dictionary, always read the whole entry. Remember that some words have more than one meaning (e.g. *form* = **1.** '(englische) (Schul-)Klasse'; **2.** 'Form'; **3.** 'Formular').
  The usage of English words is best checked in a monolingual dictionary ▶ **Skill 3**.
- Read through your translation and make sure it is consistent and makes sense. It should sound as if it were written in the target language, and not as if it has been translated.

▶ Skill 6 Working with charts and graphs

When researching a topic, you will frequently come across statistics which are presented using a chart or graph. Moreover, when giving a presentation, you may find it useful to present figures you have found in a chart or graph.

pie chart

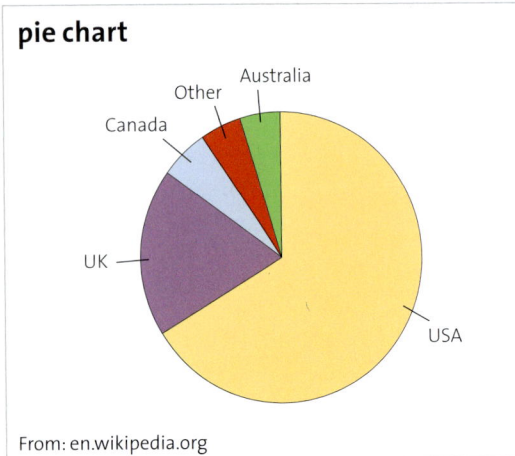

From: en.wikipedia.org

The pie chart[1] on the left shows the relative numbers of native English speakers in the major English-speaking countries of the world in 1997.

- A pie chart is divided into **slices**[2], or **sectors**, which represent parts of a whole – usually in the form of percentages of 100%. It can be an effective way of displaying information, particularly if you want to demonstrate the size of one slice as compared to the whole pie.
- The pie chart is said to be the most frequently used statistical chart in the business world and the media.

[1] **pie chart** Tortendiagramm, Kreisdiagramm   [2] **slice** Anteil, Stück

bar chart

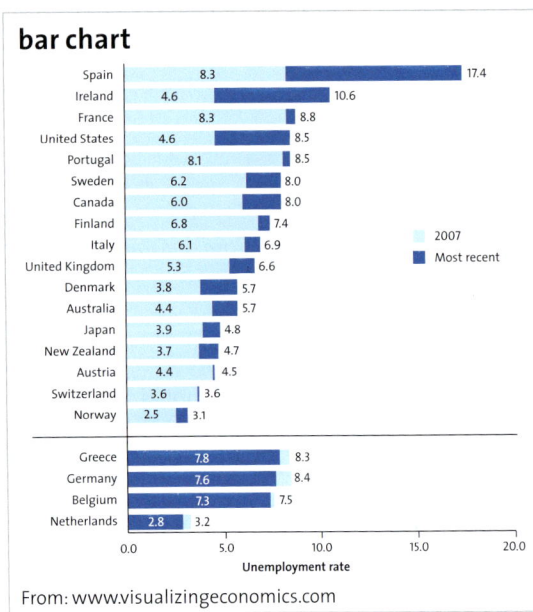

From: www.visualizingeconomics.com

The bar chart[3] on the left shows global unemployment rates and how they have developed.

- A bar chart is a chart with **rectangular**[4] **bars**, which can be either **horizontal** or **vertical**, whose lengths are proportional to the values they represent (here they are percentages).
- Bar charts are frequently used to compare two or more items, and they may also demonstrate changes over time, etc.

[3] **bar chart** Säulendiagramm, Balkendiagramm
[4] **rectangular** [rek'tæŋɡjələ] rechteckig

line graph

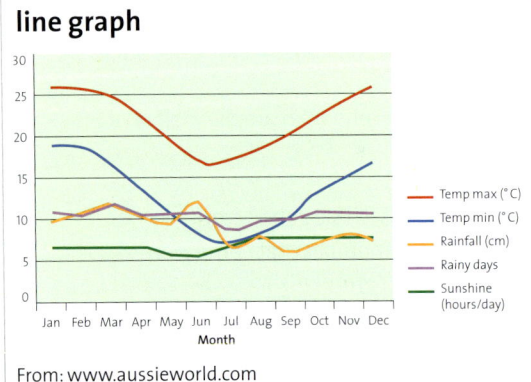

From: www.aussieworld.com

The line graph[5] on the left shows the weather conditions in Sydney, New South Wales (Australia).

- A line graph shows the relationship between numbers or items on the **vertical axis**[6] and on the **horizontal axis**. The measurements are **plotted**[7] or **charted**[8] between these two axes.
- A line graph is suitable for presenting continuity and trends.

[5] **line graph** Liniendiagramm   [6] **axis** Achse, Koordinate
[7] **plot sth.** etwas einzeichnen, etwas eintragen   [8] **chart sth.** etwas aufzeichnen, etwas darstellen

▶ Skill 7 Doing research

You will often be expected to do research on your own, for example for a presentation or a term paper.

The following steps will help you to use your time and resources effectively:

Step 1: Clarify what your topic is. Write it down in one sentence as if you were explaining it to someone. Make a list of keywords, names, etc. that you want to check on.

Step 2: Start with reference works (e.g. an encyclopedia, Wikipedia, etc.) to get an overview of your topic.

Step 3: The subject catalogue of a well-stocked library will usually lead you most directly to specialized literature. If you are working with an online catalogue or the Internet, use your keywords (Step 1) to locate relevant sources (cf. 'Using the Internet for research' below).

Step 4: Keep in mind that sources vary in their reliability (a book by a university professor is more reliable than an anonymous blog, for example) and that the presentation of the facts in a text can be very one-sided (for example, when a biotech firm defends GM foods).

Step 5: Make notes on all the information that you find relevant for your topic. Using index cards makes it easier to sort the information later.

Step 6: Note your sources carefully (author, title, place and date of publication, publisher; URL if you are using online material). You will need this information later, e.g. in a bibliography, etc.

Using the Internet for research

The Internet has become an indispensable gateway to knowledge. But since anyone can publish on the Internet, you should be particularly careful about your sources.

Another problem is the huge volume of information that can be retrieved. Finding exactly what you are looking for may be a challenge. The following tips can help you.

- Set the preferences (settings) on your search engine to limit results to English-language sources:

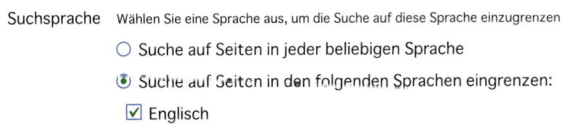

Suchsprache Wählen Sie eine Sprache aus, um die Suche auf diese Sprache einzugrenzen.
- ◯ Suche auf Seiten in jeder beliebigen Sprache
- ◉ Suche auf Seiten in den folgenden Sprachen eingrenzen:
- ☑ Englisch

- Combine headwords to get the most exact results. For example, you want to collect information on the protest movement against radioactive waste. Googling 'atomic energy' yields ca. 11,000,000 pages; entering 'atomic energy waste protest' reduces the number to ca. 4,000,000.

- Linking headwords with OR produces results with either headword:

atomic energy OR nuclear energy [Search]

- Use quotation marks to find the exact source of a quote:

"we have nothing to fear but fear itself" [Search]

- Putting a minus sign in front of a headword excludes it from the results:

Elizabethan playwrights -Shakespeare [Search]

This search request will yield information that deals mainly with other dramatists of Shakespeare's time.

▶ Skill 8 Skimming and scanning

Skimming is used to quickly identify the main ideas of a text, i.e. to get the gist. You can do this by looking at the text quickly and using all the clues to get an idea of what the text is about. People often skim when they have lots of material to read in a limited amount of time. Use skimming when you want to see if an article is relevant to your research, e.g. for a presentation or an essay.

Step 1: Look at the title, subtitle, subheading(s), and the illustrations, pictures and their captions.
Step 2: Look at the first and last sentences of each paragraph – one is usually the topic sentence (i.e. the sentence that shows what aspect is being dealt with).
Step 3: Don't read every sentence in detail; run your eyes over the whole text and look for words and phrases that seem important. It may help if you underline them.

Scanning is a technique we often use when looking up a word in the dictionary or a name in a telephone book. You can also use it to find out if a text contains specific information of interest for your topic.

Step 1: Choose a keyword or phrase to search for. In most cases, you know what you are looking for, so you are concentrating on finding a particular piece of information.
Step 2: Move your eyes quickly down the page in search of the specific word(s) or phrase(s) you have in mind.

▶ Skill 9 Reading non-fiction

When you read factual texts, you should have a set of methods to help you understand and remember better what you have to or want to read. These techniques are useful whenever you have to absorb a lot of information from non-fictional texts such as newspaper articles, articles from encyclopedias, extracts from non-fictional books, etc.

'I can't read a word of this essay of yours. Excellent work.'

Step 1: Skim the text ▶ Skill 8 to find out quickly what it is about, whether you are interested in its topic or whether it is relevant to your research, e.g. for a presentation or an essay. In addition, it might help to write down a couple of sentences.
Step 2: Examine the structure of the text to see how the information is presented. Look at the layout: paragraphs (often with sub-headings) usually deal with different aspects of the topic. Scan the text ▶ Skill 8 for adverbs, connectors or conjunctions (e.g. *only, but while, that is why, while ...*) the writer uses to structure his or her arguments.
Step 3: Extract information by reading the text carefully and slowly. Look up any words you do not know or understand (but remember you can guess the meaning of a lot of words ▶ Skill 1). Underline or copy out keywords and expressions.
Step 4: Organize your information in a way that will help you to remember it. Look at what you have underlined or at the notes you have made. Can the information be structured or organized? Make headings – perhaps on index cards – and put keywords and expressions under the appropriate headings (headings often correspond to the paragraphs in the text), or make a mind map. Now read through your information again – is anything missing; do you need to go back to the text for further details?

Step 5: Present your information to inform others. You should now be able to use your notes to summarize the text ► **Skill 11**, to talk about it or to include the information you have gained in an essay ► **Skill 10** or talk. ● **Language help 1**

Step 6: React to the text with your own opinion. State what your opinion on the topic was before you read the text. Explain whether the text has confirmed your opinion, made you rethink the issue or even changed your mind. ● **Language help 2**

Language help 1

- This text deals with ...
- The writer puts forward the view / points out that ...

Language help 2

- I always thought/believed/felt that ...
- I was unaware of the fact that ...
- It is worth considering that ...

► Skill 10 Writing an essay

When writing an essay, you are required to present a topic in a coherent way, for example by presenting arguments for and against a topic.

Step 1: Read the task carefully and pay particular attention to the special vocabulary used in tasks, so that you know exactly what is required of you.

Step 2: Proceed as you would with any other text, from first draft to final draft giving yourself enough time to re-read your essay and check for grammar and spelling mistakes, but consider the following points as well:

- In your introduction restate the problem from your task. You can also state your main points and explain the structure of your text.
- Present your arguments in the main part of your essay. Devote one paragraph to each of the arguments. Use connectors to make your text coherent. ● **Language help**
- If you are asked to analyse or examine an issue, arrange your arguments according to their importance, from the weakest to the strongest point or vice versa.
- If you are asked to discuss a topic, i.e. weigh up the pros and cons, you can take the argumentative approach and deal with all the arguments in favour of the topic first and then all the arguments against the topic next. Alternatively, you can discuss one aspect at a time, presenting ideas for and against each aspect one after the other.
- If you are asked to compare two things, deal with the similarities first and then the differences or vice versa.
- Make a concluding statement in which you sum up your arguments.

Language help

- *Stating your opinion*
  In my opinion ... / To my mind ... / I think/feel/believe ...
- *Enumerating facts, etc.:*
  a) starting
     Firstly ... / First of all ... / For one thing ...
  b) continuing
     Secondly ... / Furthermore ... / Moreover ... / Besides ... /Also ... / In addition (to) ... / As well as ... / Another point is ...
  c) finishing
     Above all ... / Finally ...
- *Giving an example*
  ... for example ... / ... for instance ... / ... such as ... / ..., say, ...
- *Pointing out reasons or consequences*
  For this reason ... / For these reasons ... / Due to / Because of ... / As a result (of) ... / Consequently ... / As a consequence ... / Therefore ... / So ... / That explains why ...
- *Emphasizing*
  In fact ... / As a matter of fact ... / In reality ...
- *Contrasting*
  On the one hand ... on the other hand ... / However ... / Nonetheless ... / At first ... but then ... / Although ... / In spite of ... / Whereas ...
- *Conceding a point*
  Of course ... / To be sure ... / Admittedly ...
- *Pointing out a restriction or objection*
  Still ... / After all ...
- *Referring to a point in time or a development*
  At that time ... / In those days ... / Eventually ... / In the long run ... / In the course of time ... / Meanwhile ... / At the same time ...
- *Coming to a conclusion*
  All in all ... / Ultimately ... / In the final analysis ... / In conclusion ... / For the reasons mentioned above ... / To sum up ... / To conclude ... / I would like to conclude by saying that ...

▶ Skill 11 Writing a summary

A summary, for example of a text, film, play or listening piece, is a structured account of all the essential information.

Step 1: Read, watch or listen to the text you are dealing with as often as possible and make sure you have understood it. If you have enough time, look up any words you don't understand ▶ Skill 3.

Step 2: Note down what the text is about in one or two sentences.

Step 3: If you are dealing with a written text, structure the text by dividing it into parts. Note down a heading for each part which shows what that particular part deals with.

Step 4: Read, watch or listen to the text again and decide what your reader needs to know about the text by asking yourself *Who? What? When? Where? Why?* Use these questions to write down keywords and to make notes. Write down only the things one needs to know in order to understand the text.

Step 5: In the first sentences of your summary give basic information about the text, e.g. the title and the name of the author, as well as the overall topic of the text. Give the main points but leave out examples or quotations. Write in the simple present. Use your own words as much as possible and don't give your opinion on the subject. Change direct speech into indirect speech.

Step 6: A summary should be no longer than about one third to one half of the original text. If you have been told to use a certain number of words in your summary, count the words to check if you are within the limits. If you need to cut down words, remember that examples and illustrations as well as figurative language are not essential in a summary. Other methods of cutting down the number of words include:

- using single words instead of long phrases;
- combining a series of short sentences into a long sentence;
- using conjunctions and clauses.

Language help

- The text presents / deals with the question of ...
- The author points out / claims / makes clear / doubts that ...
- The author compares x to y / contrasts x with y.
- The author asks the question of whether ...

Tip

If the summary is part of an **exam**, it is best to deal with all the other tasks based on the text before writing your summary, as you will have a much clearer idea of what the text is about.

Vocabulary you should learn (from pages 6–25)

Part A:
Living in Wonderland: Modern Technology

A1 Getting Connected		
p.8	(to) **approach** sb. [ə'prəʊtʃ]	(to) come towards sb.	*auf jdn. zukommen, sich jdm. nähern*
	dashboard ['dæʃbɔːd]	the inside of a car, in front of the driver, where the controls are	*Armaturenbrett*
	column ['kɒləm]	Have you read the ~ in the New York Times about climate change?	*Kolumne; Zeitungsartikel*
	(to) **relate** sth. [rɪ'leɪt]	*synonyms:* (to) tell sth. / (to) narrate sth. / (to) recount sth.	*etwas berichten, etwas erzählen*
	(to) **label** sb./sth. **(as)** sth. ['leɪbl]	Helen was ~**led** a nerd after she admitted to liking *Star Wars*.	*jdn./etwas als etwas bezeichnen*
	accessible [ək'sesəbl]	*word family:* (to) access ['ækses] – access (n) – **accessible** – accessibility	*erreichbar, leicht zugänglich*

A2 In Our Image: The Age of Robotics		
p.10	(to) **interact with** sb. [ˌɪntər'ækt]	*synonym:* (to) deal with sb.	*mit jdm. interagieren; mit jdm. umgehen*
p.11	**interface** ['ɪntəfeɪs]	the point where two things affect each other	*Schnittstelle*
	application [ˌæplɪ'keɪʃn]	*word family:* (to) apply sth. – applicable [ə'plɪkəbl, 'æplɪkəbl] – applicant – **application**	*Anwendung; Einsatzgebiet*

Words in Context:
The Scientific Revolution

p.12	**hypothesis** [haɪ'pɒθəsɪs]	My ~ is that if there were less cars, people living in cities would be healthier.	*Hypothese, Annahme*
	empirical data [ɪmˌpɪrɪkl 'deɪtə]	*other collocations with* **empirical**: ~ science / research / study / method	*empirische Daten*
	pure science [ˌpjʊə 'saɪəns]	a theoretical study of the sciences, rather than a practical study	*theoretische Wissenschaft, reine Wissenschaft*

(p.12)	**basic research** [rɪ'sɜːtʃ, 'riːsɜːtʃ]	*other collocations with* **research**: medical / historical / scientific ~	*Grundlagenforschung*
	applied science [ə‚plaɪd 'saɪəns]	the opposite of pure science	*angewandte Wissen-schaft*
	engineering [‚endʒɪ'nɪərɪŋ]	*word family:* engine – engineer *(n)* – (to) engineer sth. – **engineering**	*Ingenieurwesen; Technik*
	(to) implement new discoveries ['ɪmplɪment, dɪ'skʌvəriz]	*other collocations with* **implement**: (to) ~ a law / a plan	*Neuentdeckungen verwirklichen*
	major advance [‚meɪdʒər_əd'vɑːns]	In the past decade there have been ~ **advances** in wind technology.	*großer Fortschritt; Meilenstein*
	breakthrough ['breɪkθruː]	Scientists have recently made a ~ in cancer research.	*Durchbruch*
	progress ['prəʊgres ☆ 'prɑːgrəs]	improving or getting closer to completing something	*Fortschritt(e)*
	state-of-the-art technology [tek'nɒlədʒi]	The new space shuttle was made with **state-of-the-art ~.**	*modernste Technologie; dem neuesten Stand entsprechende Technologie*
	(to) come under fire	The Prime Minister has ~ **under fire** for his views on immigration.	*unter Beschuss geraten, Kritik ausgesetzt sein*
	ethical ['eθɪkl]	connected with beliefs and principles about what is right and wrong	*moralisch, ethisch*
	genetically modified *(abbr* **GM***)* [dʒə‚netɪkli 'mɒdɪfaɪd]	The environmental effects of ~ **modified** food are now being researched.	*genetisch verändert, genmanipuliert*
	(to) call for a moratorium [‚mɒrə'tɔːriəm]	Environmental authorities have ~**ed for a ~** on whaling.	*einen Stopp / ein Moratorium fordern*
	deployment [dɪ'plɔɪment]	I strongly object to the ~ of nuclear weapons.	*Einsatz, Entwicklung*
	long-term effect [ɪ'fekt]	Lung cancer and heart disease are some of the ~ **effects** of smoking.	*Langzeitwirkung, langfristige Folge*
	unforeseen consequence [‚ʌnfɔːsiːn 'kɒnsɪkwəns]	The industrial revolution had notable ~ **consequences**.	*ungeahnte Folge, nicht vorhersehbare Auswirkung*
	controversy ['kɒntrəvɜːsi, kən'trɒvəsi]	*synonym:* dispute	*Auseinandersetzung; Streit*
	moral objection [‚mɒrəl_əb'dʒekʃn]	*other collocations with* **moral**: ~ issue / dilemma / question / standard / sense / outrage	*moralischer Einwand*

(p.12)	(to) **tamper with nature** ['neɪtʃə]	Is cell cloning **~ing with nature**?	*der Natur ins Handwerk pfuschen; die Natur manipulieren*
	major challenge [ˌmeɪdʒə 'tʃælɪndʒ]	extremely demanding task or situation	*große Herausforderung*
	(to) **strike a balance (between** A **and** B) ['bæləns]	It's difficult to **~ a balance between** homework **and** a healthy social life.	*den Mittelweg finden, das richtige Verhältnis (zwischen A und B) finden*
	global warming ['gləʊbl]	As **global ~** continues many communities will suffer.	*globale Erwärmung, Erwärmung der Erdatmosphäre*
	depletion of natural resources [dɪ'pliːʃn, ˌnætʃrəl rɪ'sɔːsɪz]	The **~ of natural resources** such as water can lead to conflict over access to resources.	*Raubbau an natürlichen Ressourcen; massiver Abbau natürlicher Ressourcen*
	loss of biodiversity [ˌbaɪəʊdaɪ'vɜːsəti]	The new factory may be responsible for the **loss of ~** in the forest.	*Verlust der Artenvielfalt, Rückgang der Artenvielfalt*
	(to) **threaten** sb./sth. ['θretn]	The apes' habitat is being **~ed** by the logging industry.	*jdn./etwas bedrohen*
	(to) **quantify** sth. ['kwɒntɪfaɪ]	*synonyms:* (to) measure sth. / (to) scale sth.	*etwas messen, etwas quantifizieren*
	environmental impact [ɪnˌvaɪrənmentl_'ɪmpækt]	*other collocations with* **environmental:** **~** issues / problems / damage / movement	*Auswirkung auf die Umwelt*
	carbon footprint [ˌkɑːbən 'fʊtprɪnt]	You can reduce your **~ footprint** by walking instead of driving.	*CO_2-Fußabdruck, CO_2-Bilanz*
	greenhouse effect ['griːnhaʊs_ɪ,fekt]	The **~ effect** is caused by the sun's heat being trapped in the earth's lower atmosphere.	*Treibhauseffekt*
	sustainability [səˌsteɪnə'bɪləti]	*word family:* (to) sustain – sustainable – **sustainability**	*Nachhaltigkeit, nachhaltige Entwicklung*
	renewable energy* resources [rɪˌnjuːəbl_'enədʒi]	Wind and wave power are **~ energy resources**.	*erneuerbare Energiequellen*
	conservation [ˌkɒnsə'veɪʃn]	environmental protection	*(Natur-)Schutz, Erhaltung*
	(to) **rise to the challenge** ['tʃælɪndʒ]	The Olympic ice skater will **rise to the ~** and compete against the new world champion.	*sich der Herausforderung stellen, sich der Herausforderung gewachsen zeigen*
	(to) **deal with the consequences of** sth. ['kɒnsɪkwənsɪz]	You have to learn to **~ with the ~** of your actions.	*mit den Folgen von etwas umgehen*

Part B:
Cracking the Code: Genetics

B1 How Designer Children Will Work

p.14	**gene** [dʒiːn]	People say I have more of my mum's **~s** than my dad's. She had very blue eyes too.	*Gen, Erbfaktor*
	genome ['dʒiːnəʊm]	The ~ of dolphins and humans have been compared in diabetes research.	*Genom, Erbgut*
	chromosome ['krəʊməsəʊm]	the small gene carrying structure found in animal and plant cells	*Chromosom*
	uterus ['juːtərəs]	the female organ in which babies develop	*Gebärmutter*
	in vitro [ɪn 'viːtrəʊ]	Her baby was conceived **in ~**.	*künstlich; im Reagenzglas*
	cloning* *(n)* ['kləʊnɪŋ]	*collocations with* **cloning**: human / reproductive / therapeutic ~	*Klonen*
	nucleus ['njuːkliəs]	the centre of the cell where its genetic information is contained	*Zellkern*
	genetic engineering* [dʒə,netɪk_,endʒɪ'nɪərɪŋ]	*other collocations with* **genetic**: ~ code / disease / research / fingerprint	*Gentechnik, Gentechnologie*
	stem cell ['stem sel]	The medical possibilities of ~ **cell** research are exciting, yet controversial.	*Stammzelle*
	hereditary [hə'redɪtri ☆ hə'redɪteri]	biologically passed down to a child from its parents before it is born	*erblich, vererbt*

B2 Born for a Purpose

p.15	(to) **fill sb. in (on** sth.)	*synonyms:* (to) inform sb. (about/of sth.) / (to) brief sb.	*jdn. einweihen; jdn. informieren (über etwas)*
	(to) **multiply** ['mʌltɪplaɪ]	*opposite:* (to) become extinct	*sich vermehren*
	(to) **vacation** *(AE)* [və'keɪʃn, veɪ'keɪʃn]	We're **~ing** in the Hamptons this summer.	*Ferien machen, Urlaub machen*
	distinction [dɪ'stɪŋkʃn]	The student argued that there is a clear ~ between being inspired by someone and plagiarism.	*Unterschied*
	(to) **file for divorce** [dɪ'vɔːs]	*other collocations with* (to) **file for**: (to) ~ unemployment / bankruptcy	*die Scheidung einreichen*
	flattering ['flætərɪŋ]	The book's reviews were very ~.	*schmeichelhaft*

(p.15)	(to) **hook** sth. **up** [hʊk]	(to) connect sth.	*etwas verbinden*
	precious ['preʃəs]	*synonym:* treasured	*wertvoll, kostbar*
	(to) **bargain for** sth. ['bɑːgən]	The car thief got more than he **~ed for** when he turned and saw a baby in the backseat.	*etwas erwarten,* *mit etwas rechnen*

B3 GM Food – Does Anybody Want It?

p.16	**overcast** [,əʊvə'kɑːst]	*synonyms:* cloudy / clouded	*bewölkt, bedeckt, trüb*
	gloomy ['gluːmi]	She turned on a light so the room didn't look so **~**.	*düster; deprimierend*
	mutilation [,mjuːtɪ'leɪʃn]	the act of damaging sth. or injuring sb. in a violent way	*Verstümmelung;* *Verletzung*
	wrench *(AE)* [rentʃ]	a metal tool, often with an adjustable head, used for holding onto and turning things	*Schraubenschlüssel*
	test plot	an area used to grow plants for testing purposes	*Versuchsbeet*
	(to) **tread (trod, trodden)** [tred]	Don't let the cows in the garden – they might **~** on my roses.	*(zer)treten*
	debris ['debriː ☆ də'briː]	**D~** was found more than 500 m from the site of the bomb blast.	*Überbleibsel; Trümmer*
	(to) **survey** sth. [sə'veɪ]	*synonyms:* (to) examine sth. / (to) study sth. / (to) analyse sth.	*etwas begutachten*
	rash [ræʃ]	There has been a **~** of burglaries in high-profile jewellery stores.	*hier: Serie (unliebsamer Vorkommnisse)*
	breeding technique ['briːdɪŋ tek,niːk]	Scientists are using new **~ techniques** to make plants more resistant to disease.	*Zuchtmethode*
	pesticide ['pestɪsaɪd]	Organic food is grown without the use of **~**.	*Pflanzenschutzmittel,* *Pestizid*
	plentiful ['plentɪfl]	The strawberry harvest was **~** this year because of the good weather conditions.	*ertragreich, reichlich*
	crop [krɒp]	Fortunately, the **~s** survived the wet summer.	*Getreide, Feldfrucht*
	(to) **sift through** sth. [sɪft]	*synonyms:* (to) scan sth. / (to) search through sth.	*etwas durchsuchen*
	drought [draʊt]	an extended period of time with little or no rain so that water levels are very low	*Dürre, Trockenperiode*

p.17	**yield** (n) [jiːld]	Using fertilizers can result in higher **~s**.	*Ertrag*
	(to be) underway (adj)	Building work is already **~** on the new Olympic stadium.	*im Gange (sein)*
	vaccine ['væksiːn ☆ væk'siːn]	The doctor bought extra stocks of the flu **~** before winter.	*Impfstoff*
	diarrhoea (BE) / **diarrhea** (AE) [ˌdaɪə'rɪə]	Drinking unclean water can result in **~**.	*Durchfall*
	(to) inoculate sb. [ɪ'nɒkjuleɪt]	The birds will have to be **~ed** against the disease.	*jdn. impfen*
	(to) take one's cue from sb. [kjuː]	I'm **~ing my cue from** Bill – I'm going to invest in computers.	*sich nach jdm. richten; jdn. kopieren*
	(to) steer clear of sth. [stɪə]	*synonyms:* (to) avoid sth. / (to) get around sth.	*etwas meiden; einer Sache aus dem Weg gehen*
	merger ['mɜːdʒə]	Many expect the companies' **~** will mean job losses and pay cuts.	*Fusion, Zusammenschluss*

Part C:
The Challenge of Climate Change

C2 The Future of Energy – the Energy of the Future			
p.21	**circumstance** ['sɜːkəmstəns]	Aid organizations aim to improve the social **~s** of many countries.	*Umstand, Situation*
	(to be) constrained by sth. [kən'streɪnd]	I often feel **~ed** when I can't explain what I mean in a foreign language.	*von etwas eingeschränkt sein*
	(to) precede sth. [prɪ'siːd]	(to) happen or come before sth.	*einer Sache vorausgehen*
	surplus ['sɜːpləs]	*synonyms:* abundance / affluence	*Überschuss*
	available [ə'veɪləbl]	*synonyms:* at hand / existent	*verfügbar, vorhanden*
	fortunate ['fɔːtʃənət]	*collocations with* **fortunate:** **~** circumstance / coincidence / event	*glücklich, vom Glück begünstigt*
	(to) inhabit sth. [ɪn'hæbɪt]	My great-grandmother **~ed** an age in which space travel seemed a fantasy.	*in etwas leben, etwas bewohnen*
	(to) replace sth. [rɪ'pleɪs]	She had to drink lots of water during the marathon to **~** the fluid lost.	*etwas ersetzen; etwas erneuern*

(p.21)	(to) **take** sth. **for granted** ['grɑːntɪd]	Access to clean drinking water is **taken for ~** by the developed world.	*etwas als selbst-verständlich erachten*

C3 Science to the Rescue?

p.23	(to) **combat** sth. ['kɒmbæt]	*synonyms:* (to) fight sth. / (to) battle against sth.	*etwas bekämpfen*
	(to) **maintain** sth. [meɪn'teɪn]	*collocations with* **maintain**: (to) ~ jobs / silence (about sth.)	*etwas (aufrecht)erhalten*
	(to) **imply** sth. [ɪm'plaɪ]	The teacher **~ied** that if we took part in the school concert we would get better marks.	*etwas andeuten*
	deforestation [ˌdiːˌfɒrɪ'steɪʃn]	*opposite:* reforestation	*Abholzung, Entwaldung*
	(to) **enhance** sth. [ɪn'hɑːns]	*synonyms:* (to) increase sth. / (to) augment sth.	*etwas verbessern; etwas steigern*
	(to) **retain** sth. [rɪ'teɪn]	Too much salt can cause the body to ~ water.	*etwas speichern*
	(to) **amplify** sth. ['æmplɪfaɪ]	*opposite:* (to) weaken	*etwas verstärken*
	(to) **thaw** [θɔː]	As the sun came out, the ice began to ~.	*tauen, schmelzen*
	immediate effect [ɪˌmiːdiət_ɪ'fekt]	*other collocations with* **immediate**: ~ reaction / response / measure	*sofortige Wirkung*
	(to) **inflict harm** [ɪn'flɪkt]	The locals are worried about the new factory **~ing harm** on the environment, but it will bring jobs to the area.	*Schaden zufügen*

CLIMATE CHANGE CONFERENCE

NO ONE WILL TAKE ME SERIOUSLY IN THESE SHOES

CARBON FOOTPRINT

CHRIS MADDEN

Topic Vocabulary

More useful vocabulary to help you discuss the themes in this book

Science		
advance [əd'vɑːns]	Recent ~s in technology have drastically improved the speed with which we can communicate with one another.	*Fortschritt*
AIDS	'Acquired Immune Deficiency Syndrome' – a condition in which the body is unable to protect itself from infection	*AIDS*
antibiotics *(pl)* [ˌæntibaɪ'ɒtɪks]	The doctor prescribed her a course of ~ and the infection was gone in less than a week.	*Antibiotika*
breakthrough ['breɪkθruː]	*collocations with* **breakthrough**: a dramatic / major / significant ~	*Durchbruch*
cancer ['kænsə]	Five years passed and she was given the all-clear – she had finally beaten ~!	*Krebs(erkrankung)*
chemotherapy [ˌkiːməʊ'θerəpi]	a common course of treatment for cancer patients	*Chemotherapie*
diabetes [ˌdaɪə'biːtiːz]	*word family:* **diabetes** – diabetic [ˌdaɪə'betɪk] *(n, person)* – diabetic *(adj)*	*Diabetes, Zuckerkrankheit*
discovery [dɪ'skʌvəri]	*word family:* (to) discover – discoverer – **discovery**	*Entdeckung*
DNA (d**eoxyribo**n**ucleic** a**cid)**	a chemical in the cells of animals and plants carrying genetic information	*DNS (Desoxyribo-nukleinsäure)*
ethics *(pl)* ['eθɪks]	moral principles that control how one behaves in a certain context	*Ethik*
experiment [ɪk'sperɪmənt]	*word family:* (to) experiment – **experiment** *(n)* – experimental	*Experiment*
foetus *(BE)* **/ fetus** *(AE)* ['fiːtəs]	Smoking or consuming alcohol during pregnancy can harm a developing ~.	*Fetus, Fötus*
genetic disorder [dʒəˌnetɪk dɪs'ɔːdə]	*other collocations with* **disorder**: eating / mental / nervous ~	*genetische Störung*
HIV (h**uman** i**mmuno-**d**eficiency** v**irus)**	With over five million infected people, South Africa is the ~ capital of the world.	*HIV*
IVF (i**n** v**itro** f**ertilization)**	a process which fertilizes a female egg outside the woman's body before being put in her uterus to develop	*In-vitro-Befruch-tung/-Fertilisation*
laboratory [lə'bɒrətri ☆ 'læbrətɔːri]	The animal rights activists protested outside the ~.	*Labor*
(medical) treatment [ˌmedɪkl 'triːtmənt]	*synonyms:* care / therapy	*(medizinische) Behandlung*
medicine ['medsn, 'medɪsn]	a substance taken to cure an illness or relieve its symptoms	*Medizin, Medikament(e)*
(to) **operate on** sb. ['ɒpəreɪt]	*word family:* (to) **operate** – operation – operator	*jdn. operieren*

painkiller ['peɪnkɪlə]	After waking up with a headache, Ralph took a couple of ~s to be able to go to work.	*Schmerzmittel*
(to) **research** [rɪ'sɜːtʃ]	*word family:* (to) **research** – research *(n)* – researcher	*forschen*
sperm bank ['spɜːm bæŋk]	*other collocations with* **sperm**: sperm donor / sperm count	*Samenbank*
symptom ['sɪmptəm]	*synonyms:* indication / sign	*Symptom*
test-tube baby [tjuːb]	*synonym:* IVF baby	*Retortenbaby*
tumo(u)r ['tjuːmə ☆ 'tuːmər]	After months of painful headaches she went to the doctor and was told she had a brain ~.	*Tumor, Geschwulst*

Technology

(to) **crash** [kræʃ]	I lost a lot of my work when my computer ~**ed** unexpectedly.	*abstürzen (Computer)*
identity theft [aɪ'dentəti θeft]	*other collocations with* **theft**: data / petty / car ~	*Identitätsdiebstahl, -betrug*
(to) **invent** sth. [ɪn'vent]	*word family:* (to) **invent** – invention – inventive – inventor	*etwas erfinden*
keyboard ['kiːbɔːd]	a device with which one can enter information into a computer	*Tastatur*
memory stick ['meməri stɪk]	*synonyms:* flash drive / USB drive / pen drive / thumb drive *(AE)*	*USB-Stick*
monitor ['mɒnɪtə]	He spent too much time in front of his computer ~ playing computer games.	*Bildschirm, Monitor*
robotics *(sing)* [rəʊ'bɒtɪks]	the science of designing and operating robots	*Robotik, Robotertechnik*
state of the art	*opposite:* out of date / obsolete	*auf dem neuesten Stand der Technik*
Wi-fi (wireless fidelity) ['waɪfaɪ]	My favourite café had free ~ for paying customers and was popular with the students in the neighbourhood.	*WLAN (wireless local area network)*
wireless *(adj)* ['waɪələs]	transmitted through electrowaves and without a wire connection	*Funk-*

Environment

(to) **become extinct** [ɪk'stɪŋkt]	*synonym:* (to) die out	*aussterben*
carbon footprint [ˌkɑːbən 'fʊtprɪnt]	a measurement of the daily amount of carbon dioxide produced by a person or company	*CO_2-Bilanz*
climate change ['klaɪmət tʃeɪndʒ]	The extinction of mammoths has been attributed to ~ **change** rather than to hunting by humans.	*Klimawandel*

deforestation [ˌdiːfɒrɪ'steɪʃn]	*opposites:* afforestation / reforestation	*Entwaldung,* *Waldabbau*
(to) destroy the ozone layer ['əʊzəʊn ˌleɪə]	*opposite:* (to) repair the ozone layer	*die Ozonschicht* *zerstören*
emission [ɪ'mɪʃn]	The council took measures to reduce exhaust pipe ~s in the city centre.	*Ausstoß, Emission*
famine ['fæmɪn]	a lack of food in a region for a period of time	*Hungersnot*
flood [flʌd]	*opposite:* drought [draʊt]	*Überschwemmung*
fossil fuel ['fɒsl ˌfjuːəl]	fuel made up of natural materials, usually dead organisms which have decomposed over millions of years (e.g. coal or oil)	*fossiler Brennstoff*
global warming ['gləʊbl]	*other collocations with* **global**: ~ economy / finance / business	*Erderwärmung,* *Erwärmung der* *Erdatmosphäre*
greenhouse effect [ɪ'fekt]	an increase in gases like carbon dioxide around the Earth that results in trapped heat and a gradual rise in temperature	*Treibhauseffekt*
oil spill ['ɔɪl spɪl]	The ~ **spill** in the Gulf of Mexico in 2010 far exceeded any other in living memory.	*Ölpest, Ölkatastrophe*
(to) pollute sth. [pə'luːt]	*opposite:* (to) clean sth.	*etwas verschmutzen*
power station	*other collocations with* **station**: police / fire ~	*Kraftwerk*
radioactive fallout [ˌreɪdiəʊæktɪv 'fɔːlaʊt]	*another collocation with* **fallout**: political ~	*radioaktiver* *Niederschlag*
renewable energy* [rɪˌnjuːəbl_'enədʒi]	*opposite:* non-renewable energy	*erneuerbare Energie*
sewage ['suːɪdʒ]	used water and human waste, disposed of and transported away via sewers	*Abwasser*
solar panel [ˌsəʊlə 'pænl]	a flat object consisting of solar cells with the purpose of collecting energy from the sun	*Sonnenkollektor,* *Solarkollektor*
solar power (or solar energy)	*other collocations with* **power**: wind / water ~	*Sonnenenergie*
sustainable [sə'steɪnəbl]	*opposite:* unsustainable	*nachhaltig*
toxic ['tɒksɪk]	Be careful to wear gloves and a mask when handling ~ chemicals.	*giftig, toxisch*
water supply [sə'plaɪ]	The ~ **supply** has been affected by the earthquake.	*Wasserversorgung*
wildlife *(sing)* ['waɪldlaɪf]	She dedicated her life to raising awareness and conserving endangered ~.	*wild lebende Tierarten*

Selected vocabulary from the accompanying audios and videos to assist comprehension

p.10	**A2** `CD 02` In Our Image: The Age of Robotics i) The Three Laws of Robotics	

(to) **coin** sth. [kɔɪn]	(to) make up a new word or phrase that other people then start to use
credit (for sth.) ['kredɪt]	praise or approval for sth. you did
thrill [θrɪl]	strong feeling of excitement; an experience that gives you that feeling
lever ['liːvə ☆ 'levər]	Hebel
(to) **strike (struck, struck)** sb. **as** sth.	(to) give sb. a particular impression; (to) seem to have a particular quality

p.14	**B1** `DVD` How Designer Children Will Work Task 2: Collecting information from a film	

(to) **devastate** sth. ['devəsteɪt]	(to) destroy sth.
remote *(adj)* [rɪ'məʊt]	*(here)* very small
aspiring *(adj before n)* [ə'spaɪərɪŋ]	wanting to start the activity that is mentioned, e.g. *aspiring writers*
fertilization [,fɜːtəlaɪ'zeɪʃn ☆ ,fɜːrtələ'zeɪʃn]	Befruchtung
(to) **outweigh** sth. [,aʊt'weɪ]	(to) be more important than sth. else
(to) **meddle with** sth.	(to) try to change or influence sth. which is not your responsibility
shot *(infml)* [ʃɒt]	*(here)* injection
from the outset	from the beginning
(to) **go ahead**	(to) proceed; (to) begin to do sth.
insurance coverage *(AE)* = *(BE)* **insurance cover** [ɪn'ʃʊərəns ,kʌvərɪdʒ]	protection that an insurance company provides by promising to pay you money if a particular event happens
elective *(adj, fml)* [ɪ'lektɪv]	*(here)* optional
allure *(fml)* [ə'lʊə]	quality of being attractive and exciting
susceptible to sth. [sə'septəbl]	likely to be influenced or affected by sth.

(p.14)	**colon** ['kəʊlən]	Dickdarm
	the/a slippery slope ['slɪpəri]	an action that cannot be stopped easily once it has begun and can lead to serious problems
	murky ['mɜːki]	complicated and unpleasant
	concerted *(adj)* [kən'sɜːtɪd]	done in a planned and determined way, especially by more than one person, government, etc.

p.16 | **B2** `CD 03–04` Born for a Purpose **Task 4: An interview with the author**

Part 1

(to) hire sb. ['haɪə]	(to) employ sb. for a short time to do a particular job
germ of sth. [dʒɜːm]	early stage of the development of sth.
eugenics [juː'dʒenɪks]	study of methods to improve the mental and physical characteristics of the human race by choosing who may become parents
(to) wind up doing sth. *(infml)* **(wound, wound)** [waɪnd, waʊnd]	*(of a person)* (to) find yourself in a particular situation or place because of other things that have happened
cord = umbilical cord [ʌmˌbɪlɪkl 'kɔːd]	Nabelschnur
(to) push the envelope *(infml)*	(to) go beyond the limits of what is allowed to do or thought to be possible
(to) tap sb. **for** sth.	(to) make sb. give you sth.
benign [bɪ'naɪn]	(of tumours) not dangerous or likely to cause death
(to) get rid of sth.	(to) free yourself of sth. that is annoying you or that is unwanted
(to) retain sth. [rɪ'teɪn]	(to) continue to keep or contain sth.
attorney *(AE)* [ə'tɜːni]	lawyer
Guardian ad Litem *(Latin)*	Rechtsbeistand für einen Minderjährigen für die Dauer eines Prozesses

Part 2

(to) walk a tightrope ['taɪtrəʊp]	(to) be in a difficult situation in which you must act very carefully in order to be successful
suffice it to say (that) [sə'faɪs]	used to suggest that although you could say more, what you do say will be enough to explain what you mean
(to) ring true/honest ['ɒnɪst]	(to) give the impression of being true/honest

(p.16)	**precociousness** [prɪˈkəʊʃəsnəs] **= precocity** [prɪˈkɒsəti]	Frühreife, Altklugheit
	(to) whine [waɪn]	(to) complain in an annoying voice
	villain [ˈvɪlən]	bad person
	(to) blame sb. (for sth.)	(to) think or say that sb. is responsible for sth. bad
	(to) be wrapped up in sth. [ræpt]	(to) be absorbed in (doing) sth.
	drained [dreɪnd]	very tired and without energy
	(to) cringe [krɪndʒ]	(to) feel very uncomfortable about sth.

<p.20>

C1 **DVD** Take AIM at Climate Change	
Verse 1	
(to) disperse [dɪˈspɜːs]	(to) move apart and go away in different directions
Verse 2	
catch *(n)*	difficulty
(to) dwell *(fml)*	(to) live somewhere
Verse 3	
(to) mitigate sth. *(fml)* [ˈmɪtɪɡeɪt]	(to) make sth. less harmful, serious, etc.

Megan and Morag – the world's first cloned sheep aged nine months

cloning

the creation of genetically identical copies of living matter. Reproductive cloning is used to produce 'identical twins' of animals through genetic manipulation. Therapeutic cloning holds the possibility of generating body-identical cells to repair damaged human tissue through the use of stem cells. Because these are obtained from human embryos, stem cell research is strictly regulated in most countries.

embryo screening

cf. pre-implantation genetic diagnosis

genetic engineering

the science of changing the information in the genes of a living thing, usually to make it stronger or healthier. Genetic engineering makes it possible to develop plants with qualities that might never have occurred in nature. Characteristics can be transferred from one species to another, even from animal to plant and vice versa. Plant breeders have been changing the genetic make-up of plants for thousands of years. The result is that today none of the plants grown as food crops in the world would exist without human interference. Critics fear that introducing 'unnatural' life forms may cause uncontrollable damage to the environment. Plants present a particular problem, as their genetic material is transmitted by pollen, which is difficult to contain. Genetic modifications could then be spread around the world. Also, it is often claimed that large biotech firms are more interested in gaining control over world food production by making farmers dependent on their products than in reducing hunger in the world.

geo-engineering

methods that deliberately manipulate the Earth's climate to counteract the effects of global warming. Some geo-engineering techniques include carbon dioxide air capture and ocean iron fertilization. To date, no large-scale geo-engineering projects have been undertaken. Some limited tree planting and cool-roof projects are already underway, and ocean iron fertilization is at an advanced stage of research.

Kyoto Protocol [kiˈəʊtəʊ]

a protocol aimed at fighting global warming. It was initially adopted in 1997 in Kyoto, Japan, but was put into practice in 2005. Industrialized countries committed themselves to a reduction of four greenhouse gases (carbon dioxide, methane, nitrous oxide, sulphur hexafluoride) and two groups of gases (hydrofluorocarbons and perfluorocarbons) by 5.2% from the 1990 level.

Kyoto 2

an alternative framework for a new climate agreement intended to replace the Kyoto Protocol. Its aims are to stabilize greenhouse gases in the atmosphere at a level that would prevent interference with the climate system, while addressing the needs of poor countries.

traditional
light bulb

energy-saving
light bulb

light bulb (*also* bulb)

glass object that fits into an electric lamp and produces light when it is switched on. The first successful light bulb was developed by Thomas Edison between 1878 and 1880. Before that time, gas was the main source of artificial light. Light bulbs are powered by electricity of which only a small percentage is used to create light. The rest comes off as heat. For this reason many countries have begun to phase out traditional light bulbs and replace them with more energy-efficient electric lights.

Organization for Economic Co-Operation and Development (OECD)

an international organization made up of 34 countries that contribute to the development of the world economy by supporting sustainable economic growth, boosting employment and assisting growth in world trade and financial stability. It is one of the world's largest sources of economic and social data and is committed to democracy and the market economy. The OECD began in 1947 as the Organization for European Economic Cooperation (OEEC), which was founded to manage American and Canadian aid in Europe under the Marshall Plan after World War II. In 1961 the OEEC was replaced by the OECD. It has its headquarters in Paris.

pre-implantation genetic diagnosis (PGD) (*also referred to as* embryo screening)

a technique used to identify genetic defects in embryos through in vitro fertilization (IVF) before pregnancy. PGD was considered illegal in Germany under the Embryo Protection Act of 1990 until the Federal Court ruled in July 2010 that PGD could be legally used alongside reproductive medicine (in a case similar to that of the Kingsbury family in the video on p. 14). In the UK, it is permitted in certain cases in clinics that are licensed by the Human Fertilisation and Embryology Authority. In the USA as well as in several European countries, there are little or no laws limiting the use of PGD. The most common use of the technology is for screening out embryos that carry a potentially deadly gene. Recently, there has been some interest, especially in the USA, in using PGD to choose the baby's gender.

renewable energy

type of energy generated from natural sources that are naturally replenished (renewed): e.g. sunlight, wind, rain, tides or geothermal heat. There are fundamental differences between renewable energy and fossil fuels. The production of oil, coal, or natural gas fuel requires a great deal of complex equipment, physical and chemical processes, while alternative energy can be produced with basic equipment and processes. Wood, the most renewable and available energy, actually releases the same amount of carbon when burnt as it would if it degraded[1] naturally, but, of course, the time frame is very different: when wood burns, it emits carbon in one go, while degradation takes place gradually.

[1] **degrade** zerfallen

stylistic devices
methods and techniques used to produce a particular effect in a text and on the reader. Below are some common examples.

alliteration [ə͵lɪtə'reɪʃn]: the repetition of a consonant at the beginning of neighbouring words or of stressed syllables within such words to produce a rhythmic effect
Example: *Around the rugged rock the ragged rascal ran.*

allusion: the direct or indirect reference to something or somebody the reader or listener is supposed to recognize and respond to. An allusion may be to a work of literature, a historical event, a well-known person, etc.

antithesis [æn'tɪθəsɪs]: an idea that is the opposite of an idea (thesis) already put forward by a writer. Often the writer will put forward the antithesis in order to stress his own thesis.

contrast ['kɒntrɑːst ☆ 'kɑːntræst]: the bringing together of opposing views, words or characters to emphasize their difference and usually to highlight one of the opposing elements
▸ juxtaposition

irony ['aɪrəni]: saying the opposite of what you actually mean
Example: *Oh, what a nice present!* when you actually mean 'It is rather ugly'.

juxtaposition [͵dʒʌkstəpə'zɪʃn]: a very strong ▸ contrast of opposing ideas, arguments, views, mostly introduced by words like *but, however, nevertheless*

metaphor ['metəfə ☆ -fɔːr]: a comparison between two things which are basically quite unlike one another without using the words *as* or *like* (▸ simile). It is meant to create a picture (image) in your mind that sheds more light on a topic.
Example: *There's daggers in men's smiles.* (William Shakespeare, *Macbeth*)

paradox ['pærədɒks ☆ -dɑːks]: a statement that seems impossible because it contains two opposing ideas that are both true
Example: *In this rich country, there is a lot of poverty.* accumulation

rhetorical question: a question to which the answer seems obvious and is therefore not necessary. A rhetorical question pushes the reader or listener to a certain conclusion.

simile ['sɪməli]: a comparison between two things that are not really like each other. Similes use the words *like* or *as*.
Example: *My love is like a red, red rose.*
(Robert Burns, 1794)

symbol ['sɪmbl]: a thing, word or phrase signifying something concrete that stands not only for itself but also for a certain abstract idea. As in the case of a ▸ metaphor or a ▸ simile the meaning of a symbol goes beyond the literal meaning.
Example: *A red rose is often a symbol of love.*

'I didn't feel answers were necessary.
All the questions seemed rhetorical.'

PERSONS

Asimov ['æsɪmɒv]**, Isaac** (1920–1992)
was born in the Soviet Union. When he was three, his Jewish parents emigrated to the USA. Asimov grew up in Brooklyn, New York, and went on to study Chemistry at Columbia University, New York. He wrote a large number of fictional and non-fictional books dealing with science and technology, but is best known for his science-fiction novels. Together with Arthur C. Clarke and Robert Heinlein, Asimov is generally regarded as one of the major writers of science fiction in the 20th century.

Breazeal, Cynthia (born 1967)
is a professor at the Massachusetts Institute of Technology (MIT) who is well known for her work in Human-Robot Interaction. For her thesis on social exchanges between humans and robots in the late 1990s Breazeal developed the robot Kismet, which is one of the most famous humanoid robots in the world. Kismet, who is now on display at the MIT museum, demonstrates simulated human emotions and expressions, and can interact with human beings.

Friedman, Thomas L. (born 1953)
has worked as a journalist for *The New York Times* since 1981. He has written extensively on international affairs, the world economy and the Middle East. Friedman has received three Pulitzer Prizes for his work and is the author of several best-selling books on globalization and related topics. His twice-weekly column in *The New York Times* is syndicated to ca. 100 newspapers worldwide.

Monbiot ['mɒnbiəʊ]**, George** (born 1963)
is an English writer, known for his environmental and political activism. He writes a weekly column for *The Guardian*, and is the author of a number of books.

Picoult [piːˈkəʊ]**, Jodi** (born 1966)
is an American author of a number of best-selling works of young adult fiction. She is married and lives in New Hampshire with her husband and three children.

Tickell, Oliver (born 1955)
is a British environmental campaigner. He wrote the ▶ Kyoto 2 climate initiative, which functions as the foundation of a new climate agreement to replace the ▶ Kyoto Protocol after 2012. Kyoto 2 has been released as a book and is supported by many British journalists and some members of parliament. Tickell's particular focus is on stabilizing the world's greenhouse gas emissions whilst taking into account the needs of developing countries. He has also been the Green Party candidate in elections in Oxford three times.

Acknowledgements

Texts

p.8: from The New York Times, 1 November 2006 © 2011 The New York Times. All rights reserved. Used by permission and protected by the Copyright Laws of the United States. The printing, copying, redistribution, or retransmission of the Material without express written permission is prohibited; **pp.10–11:** © 1990–2005 The Chedd-Angier Production Company, Inc. All rights reserved; **p.15:** © Jodi Picoult, printed by permission of Simon & Schuster, Inc; **pp.16–17:** © held by the National Center for Case Study Teaching in Science (NCCSTS), University at Buffalo, State University of New York. Used with permission. Excerpt as provided by law, this material may not be further reproduced, distributed, transmitted, modified, adapted, performed, displayed, published, or sold in whole or in part, without prior written permission form NCCSTS; **p.21:** © 2006 George Monbiot; **pp.23–24:** © Oliver Tickell, Guardian News & Media Ltd 2009; **p.32:** © Oxford University Press 2011.

Photos/Illustrations

p.6: © Tony Auth; **pp.6–7:** Shutterstock images / © Bruce Rolff (backdrop); **p.7:** A: © Kulkafoto / Matthias Kulka, B: dieKLEINERT.de / © Martin Peschkes, C: iStockphoto / © Cristian Matei, D: Shutterstock images / © zentilia, E: iStockphoto / © Vadim Subbotin, F: picture-alliance / © dpa-Zentralbild / Jan Woitas; **p.8:** iStockphoto / © Marcello Bortolino; **p.9:** © Uwe Kraft Fotografie, Düsseldorf; **p.10:** top: Fotex: © Friedrich Sauer, middle: Gettyimages: Hulton Archive / © Frank Capri, bottom: Gettyimages: © Amy Sussman; **p.11:** Gettyimages: © Javier Pierini; **p.12:** © Jürgen Lösel, Dresden; **p.14:** © Reprinted courtesy of HowStuffWorks.com / Zeichnung: Gabriele Heinisch, Berlin; **p.15:** Agentur Focus, Hamburg: Photo Researchers / © Victor Habbick Visions; **p.17:** Alamy: © Nick Gregory; **pp.18–19:** Shutterstock images / © Luis Francisco Cordero; **p.20:** Shutterstock.com / PaulB70 (polar bear), A: Agentur Focus, Hamburg: Science Photo Library / © Ted Kinsman, B: © Hollandse Hoogte / laif, C: Alamy: Eye Ubiquitous / © Sean Aidan; **p.21:** www.cartoon Stock.com: © Vahan Shirvanian; **p.22:** charts: © Energy information Administration (EiA), bottom: Shutterstock images / © Andrew Orlemann; **p.23:** iStockphoto / © Kutay Tanir; **p.25:** © University of Houston (pie chart), Shutterstock images / © Michael Sobers (footprint); **p.26:** Shutterstock images / © Julien Tromeur; **p.27:** iStockphoto / © hundreddays; **p.28:** Shutterstock images / © FikMik; **p.29:** Alamy: © Global Warming Images; **p.30:** left: Shutterstock images / © Sergei Butorin, middle left: Alamy: © imagebroker, middle right: Shutterstock images / © Yvan, right: Alamy: © tbkmedia.de; **p.32:** www.cartoonStock.com: © Fran; **p.36:** www.cartoonStock.com: © Ron Morgan; **p.38:** © PhotoAlto / Matthieu Spohn; **p.45:** www.cartoonStock.com: © Chris Madden; **p.52:** © King-Holmes / SLP / Agentur Focus; **p.53:** top left: Shutterstock images / © Vitaly Korovin, top right: Shutterstock images / © Arkady, middle: © OECD, bottom: Shutterstock images / © Carlo Taccari; **p.54:** www.cartoonStock.com: © Mike Baldwin; **p.55:** Asimov: Gettyimages: Hulton Archive / © Frank Capri, Breazeal: Gettyimages: © Amy Sussman, Friedman: ullstein bild / © Schleyer, Monbiot: © Pascal Saez / Writer Pictures, Picoult: Gettyimages: © David Levenson, Tickell: © Oliver Tickell / www.kyoto2.org.

Cover: Shutterstock images / © Andrea Danti